METHUEN'S MONOGRAPHS
ON CHEMICAL SUBJECTS

General Editors
R. P. Bell, M.A., LL.D., D.TECH., F.R.S.
N. N. Greenwood, D.SC., PH.D., SC.D.
R. O. C. Norman, M.A., D.PHIL.

ACIDS AND
BASES

Acids and Bases

THEIR QUANTITATIVE BEHAVIOUR

R. P. Bell, F.R.S.

*Professor of Chemistry in
the University of Stirling*

METHUEN AND CO. LTD.
11 NEW FETTER LANE, LONDON EC4

© R. P. Bell, 1969
First published May 15th 1952
Reprinted, with minor corrections, 1956 and 1961
Reprinted 1964
Second edition 1969
Printed in Great Britain by
Butler & Tanner Ltd, Frome and London

SBN 416 14660 0

Distributed in U.S.A. by
Barnes & Noble, Inc.

Contents

Contents

Preface to First Edition

The subject of acids and bases has applications to a wide variety of chemical problems, and this book is an attempt to give a brief and unified account of some of these applications. The first two chapters are elementary in nature and cover material which is presented in many elementary books. It is hoped, however, that the treatment given here may be useful in showing the essential unity of a number of topics which are commonly dealt with under a diversity of headings. Chapter 4, on the interionic theory, is not necessary for understanding the remainder of the book, though the subject is an important one for any exact interpretation of acid-base phenomena.

The modified use of the word acid, due to G. N. Lewis and his school, is not introduced until Chapter 8, where it is discussed briefly. Many may regard this as an unduly conservative policy, but it arises naturally from the emphasis given in this book to the quantitative aspects of acid-base behaviour. Little is known of the quantitative behaviour of Lewis acids, but it is already clear that they do not obey the simple general relations which govern the reactions of proton acids.

No attempt has been made to give full literature references. Those inserted in the text refer to general topics rather than to single papers, and give only the authors' names, together with an approximate date. This information should be sufficient for identifying individual papers in abstracts or indexes. The general references at the end of each chapter are to books or review articles suitable for further reading.

Preface to Second Edition

Although the main plan of the book follows that of the first edition, recent developments have necessitated several major changes. In particular, the development of experimental methods for studying fast reactions has made it possible to measure directly the rates of many acid-base reactions, and this subject is discussed together with acid-base catalysis in Chapter 6, which has been completely re-written. The use of the isotopes of hydrogen for investigating acid-base reactions is now widespread, and Chapter 7 has been added to deal with this subject. The final chapter has been expanded to include a brief account of the concept of hard and soft acids and bases, which has recently extended the usefulness of the Lewis acid-base definition. Finally, the opportunity has been taken to correct minor errors, and to bring certain sections up to date, especially in the chapter on non-aqueous solvents.

R. P. B.

CHAPTER 1
The Nature of Acids and Bases

The word 'acid' is derived from the sour taste associated with this class of substance (cf. the Latin *acidus* – sour, *acetum* – vinegar) and is still used as an adjective in this sense. Other properties were included in the characteristics of an acid at an early date: for example, in 1663 Robert Boyle stated that acids had a high solvent power, would give a red colour to blue vegetable colouring matters such as litmus, and would precipitate sulphur from liver of sulphur (obtained by dissolving sulphur in potash). The antithesis to 'acid' was at first 'alkali' (rather than base), a word derived from the Arabic *al kali*, the ashes of a plant. Alkalis were regarded as having some positive properties, such as detergent power, soapy feel, and the ability to dissolve oils and sulphur, but their chief characteristic was their ability to destroy or reverse the action of acids. For example, they will react with acids to form salts, which lack the characteristic properties of both acids and alkalis, and they will restore to its original colour litmus which has been reddened by an acid.

The formation of salts from an acid and an alkali led to the view that a salt could always be regarded as being derived from two constituents of opposite natures, even when these were not known as separate substances. Sometimes the non-acidic constituent was known, but did not possess any of the typically alkaline properties. This was true, for example, of the oxides or hydroxides of the heavy metals, and in these cases it became usual to use the term 'base' rather than alkali for the non-acidic constituent of a salt. Thus in 1774 Rouelle defined a base as any substance which reacts with an acid to form a salt, and this practical definition is still the one by which the term 'base' is usually introduced into elementary chemistry.

These early views did not involve any explanation or interpretation of the observed phenomena, and the first theory of acidic behaviour which is comprehensible in modern terms was produced by Lavoisier about 1770–80. His experiments showed that several elements, such as carbon, nitrogen, and sulphur, burned in oxygen to give compounds yielding acids with water, and this led him to conclude that 'oxygen is an element common to all acids, and the presence of oxygen constitutes or produces their acidity'. This oxygen theory of acidity has left its mark on the name still used for the element (from the Greek *oxus* – sour, *gennao* – I produce) but although it was maintained by some chemists (notably Berzelius and Gay-Lussac) up to about 1840, it was assailed by contrary evidence almost from birth. It was soon realized that many oxides, such as those of sodium, potassium, and calcium, give alkalis rather than acids with water, which accorded ill with Lavoisier's theory. Further, as knowledge of the composition of compounds became more extensive and more certain it became clear that many acids contained no oxygen. Thus as early as 1787 Berthollet produced good evidence that prussic acid and sulphuretted hydrogen contained no oxgyen, and one of the most conclusive steps was Humphry Davy's well-known demonstration (1810–11) that chlorine was an element, and not a compound of oxygen, and hence that hydrochloric acid was a compound of hydrogen and chlorine only, and contained no oxygen.

Davy at first expressed the opinion that 'Acidity does not depend upon any particular elementary substance, but upon peculiar arrangement of various substances', a view which has much in common with the later developments in the concept of acids discussed in the last chapter of this book. However, it soon became clear that all the substances commonly accepted as acids did contain *hydrogen*, and Davy soon recognized hydrogen as the essential element in an acid. On the other hand many hydrogen compounds were known which had no acidic properties, and some further qualification was necessary. Thus Liebig in 1838 defined acids as 'compounds containing hydrogen, in which the hydrogen can be replaced by metals', a definition which held the field until the advent of Arrhenius's dissociation theory. It should be noted that this definition denies the name acid to the acidic oxides themselves, although their compounds with water are often ill-defined or unknown (e.g. carbonic acid, silicic acid). Bases were still regarded as substances which neutralized acids with the formation of salts, and there was no theory as to their constitution corresponding to the hydrogen theory of acids.

The definition of acids in terms of replaceable hydrogen gives no clue as to why this kind of combined hydrogen produces the characteristic properties of acids, and the only criterion which it supplies of the relative strengths of acids and bases is the often misleading one of the replacement of one acid or base by another. It was found at an early date that the catalytic effect produced by different acids showed a general correlation with the generally accepted order of their strengths, but no quantitative understanding of this was possible before the advent of the *electrolytic dissociation theory*. This theory was developed about 1880–90, largely by Ostwald and Arrhenius, and one of its most striking successes was in the field of acids and bases. It was found that the combined hydrogen atoms which give rise to acidic properties are just those which produce *hydrogen ions* in solution, and since the degree of ionization could be determined by measurements of electrical conductivity this provided a ready means of making a quantitative assessment of the strength of an acid. The mobility of the hydrogen ion is very high, and the much smaller mobilities of anions do not vary greatly among themselves: hence the concentration of hydrogen ions in a solution of an acid is roughly proportional to its conductivity. If the catalytic power of such solutions is also due to their content of hydrogen ions, then there should exist a parallelism between the catalytic activity of solutions of acids on the one hand, and their conductivity on the other. An example of such parallelism is given in Table 1, taken from data collected by Ostwald in 1884. All the figures refer to normal solutions of the acids, and the value for hydrochloric acid is taken arbitrarily as 100 in each case. The parallelism is striking, and leaves no doubt that the hydrogen-ion concentration is the controlling factor throughout. Similar regularities are found for solutions of acids in alcohols, and Table 1B shows some typical data (taken from measurements by Goldschmidt in 1895), in this case for N/10 solutions of the acids.

The dissociation theory gave a similar account of the properties of bases in terms of the production of *hydroxide ions* in solution, though the opportunities for quantitative comparison were fewer than for acids. In particular, the power of bases to neutralize acids was explained in terms of the reaction $H^+ + OH^- \rightarrow H_2O$.

Table 1A

Relative conductivities and catalytic effects of acids in
aqueous solution

Acid	Conductivity	Catalytic effect	
		(a) in the hydrolysis of methyl acetate	(b) in the inversion of cane sugar
Hydrochloric	100	100	100
Hydrobromic	101	98	111
Nitric	99·6	92	100
Sulphuric	65·1	73·9	73·2
Trichloroacetic	62·3	68·2	75·4
Dichloroacetic	25·3	23·0	27·1
Oxalic	19·7	17·6	18·6
Monochloroacetic	4·90	4·30	4·84
Formic	1·67	1·30	1·53
Lactic	1·04	0·90	1·07
Acetic	0·424	0·345	0·400
Isobutyric	0·311	0·286	0·335

Table 1B

Relative conductivities and catalytic effects of acids in ethanol

Acid	Conductivity	Catalytic effect in the esterification of formic acid
Hydrochloric	100	100
Picric	10·4	10·3
Trichloroacetic	1·00	1·04
Trichlorobutyric	0·35	0·30
Dichloroacetic	0·22	0·18

Comparisons like the above served to lay the foundation of the dissociation theory of acids and bases, but its greatest successes were in the many quantitative predictions obtained by applying the law of mass action to dissociation equilibria: these will be described in Chapter 2. The success of these quantitative developments served to mask some logical weaknesses in the definitions of acids and bases, which were regarded as substances giving rise to hydrogen and hydroxide ions respectively in aqueous solution. For example, it was not clear whether a pure non-conducting substance like hydrogen chloride should be called an acid, or whether it became one only in contact with water or a similar medium. There was also a more serious ambiguity in the definition of bases. Most of the substances which would neutralize acids belong to one of two classes: metallic hydroxides on the one hand, and amines (including ammonia) on the other. Of these, only the former could be said to 'split off' hydroxide ions (in the same sense as acids were believed to split off hydrogen ions), though substances of the latter class also produced hydroxide ions when dissolved in water. This led to the formulation of reaction schemes such as

$$NH_3 + H_2O \rightleftharpoons NH_4OH \rightleftharpoons NH_4^+ + OH^-,$$

and there was much dispute as to whether NH_3 or NH_4OH (for the existence of which there is little direct evidence) should be regarded as a base. No real decision was reached on this point, and in some quarters a distinction was made between 'anhydro-bases' like NH_3, which neutralize acids by picking up a hydrogen ion, and 'aquo-bases' like KOH which liberate a molecule of water in the process.

These ambiguities of nomenclature are not in themselves important, and any misunderstandings could easily be overcome by common sense. However, in the study of acids and bases the development of new definitions has gone hand in hand with new investigations and a better understanding of the processes involved, so that the question of exact verbal definition has more scientific interest than is usually the case. One of the first fields to show clearly the limitations of the classical definitions was that of *non-aqueous solvents*. For example, if sodium hydroxide is dissolved in ethanol, the solution will contain hydroxyl ions, just as in water. On the other hand, a solution with still stronger basic properties is obtained by dissolving sodium ethoxide in ethanol; this solution contains ethoxide ions (EtO^-) in place of hydroxide ions, and the same ion is produced in small quantity when amines are dissolved in ethanol. In ethanol it would therefore seem more natural to define bases in terms of the ion OEt^- rather than in terms of OH^-. A similar situation arises in other non-aqueous solvents: for example, in liquid ammonia the strongest bases are the metallic amides, and the characteristic basic ion is NH_2^-.

It might be thought that these complications do not apply to acids, which all contain a hydrogen atom capable of forming a hydrogen ion, but this point of view becomes untenable when we consider the *nature of the hydrogen ion in solution*. It was originally thought that this ion was correctly represented as H^+ in any solvent, its small size being made responsible for its high mobility and catalytic power. However, evidence has gradually accumulated to show that the hydrogen ion in solution is invariably solvated, and that no measurable concentration of free H^+ can be present. The proton, H^+, is unique among cations in having no electrons at all, and its effective radius will be about 10^{-13} cm compared with 10^{-8} cm for other ions. The electric field in its neighbourhood will therefore be extremely intense, and this means that a proton will have a great affinity for other molecules, especially those with unshared electrons. In fact, approximate calculations show that the union of a proton with a water molecule will be exothermic to the extent of 200,000–300,000 calories per gram-molecule, the value derived from spectra and other experimental data being 288,000 calories. The fraction of protons which remain uncombined will be given roughly by $e^{-E/RT}$, where E is the exothermicity: this fraction is therefore of the order of magnitude 10^{-210}, which is equivalent to saying that the free proton does not exist at all in aqueous solution. The same conclusion will hold in presence of any molecules with unshared pairs of electrons, while solvents not in this class (e.g. the hydrocarbons) do not normally give conducting solutions even with the strongest acids. Hence although the bare proton can be produced in a discharge tube or in nuclear reactions, and can exist in gaseous systems at very low pressures, it certainly cannot be responsible for the characteristic properties of acids in solution.

If each proton is bound to one water molecule, the resulting cation has the formula OH_3^+, and this species is known variously as the *oxonium*, *hydroxonium*, or *hydronium* ion: we shall use the last of these names. There is good reason to believe that the hydrogen ion in aqueous solution can in fact be given this formula.

The structure OH_3^+ can be written with three covalent bonds and an octet of electrons surrounding the oxygen, showing its close analogy with the ammonium ion NH_4^+. Similar structures, with three groups attached to a positively charged oxygen atom, are met with in the organic oxonium salts derived from ethers and other organic compounds containing oxygen. The analogy with the ammonium ion is shown by the properties of the very stable monohydrate of perchloric acid: this forms crystals with an ionic lattice which are isomorphous with ammonium perchlorate $NH_4^+ . ClO_4^-$, and are thus probably hydronium perchlorate, $OH_3^+ . ClO_4^-$.

The monohydrates of other strong acids can be formulated similarly, for example $OH_3^+ . NO_3^-$ and $OH_3^+ . HSO_4^-$. Moreover, there is now ample direct evidence that these solid hydrates do contain the species OH_3^+, in which three equivalent hydrogens are bound to an oxygen. The alternative formulation, $H_2O \ldots HX$, in which a water molecule is attached to an acid molecule HX by means of a hydrogen bond, can be definitely excluded. The earliest direct evidence came from observations of proton magnetic resonance spectra (R. E. Richards and J. A. S. Smith, 1951), which can distinguish clearly between three equivalent protons in an equilateral triangle and any less symmetrical arrangement. The same conclusion was reached a little later (C. C. Ferriso and D. F. Hornig, 1955) by examining the infra-red spectra of the same solids, and it was also possible to show that OH_3^+ has a pyramidal rather than a planar structure: this is what would be expected for a species having an unshared pair of electrons.

It is reasonable to suppose that the ion OH_3^+ will retain its individuality in aqueous solutions of acids, but it has proved difficult to obtain direct evidence for its presence, since many of its characteristics (for example, spectral frequencies) will resemble those of water molecules, present in large excess, and may also be obscured by the very rapid interchange of protons between H_2O and OH_3^+: it has been estimated that OH_3^+ has an average life-time of only about 10^{-13} sec in aqueous solution. However, M. Falk and P. A. Giguère (1957) have been able to detect in aqueous solutions of a number of acids infra-red absorption bands whose frequencies correspond closely to those attributed to OH_3^+ in the solid state, and there are also a number of less direct lines of evidence. The existence of OH_3^+ is also clearly demonstrated by some experiments carried out with liquid sulphur dioxide. This solvent (at $-30°$) dissolves only very small quantities of water, but dissolves hydrogen bromide to give a non-conducting solution. The solution of hydrogen bromide in sulphur dioxide will now dissolve exactly one molecule of water per molecule of hydrogen bromide, giving a highly conducting solution. These observations are readily accounted for by the reaction $HBr + H_2O \rightarrow Br^- + OH_3^+$.

The ion OH_3^+ in aqueous solution will, like any other cation, be solvated further by the looser attachment of additional molecules of water, which will be particularly favoured by the formation of hydrogen bonds. There is in fact good evidence for the existence in aqueous solutions of acids of a fairly stable ion $H_9O_4^+$ having the structure

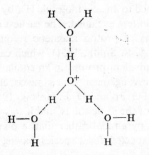

(where the broken lines represent hydrogen bonds) and it is interesting to note that both OH_3^+ and $H_9O_4^+$ can be observed in the mass spectrometer when mixtures of water vapour and hydrogen are bombarded with electrons. However, the proton binds the first molecule of water much more strongly than it does the succeeding three, and it is adequate for most purposes to represent the hydrogen ion in aqueous solution as OH_3^+. The omission of any further solvation is in line with the usual practice for metallic cations: for example, it is usual to write Li^+ rather than $Li(H_2O)_4^+$ for the lithium ion in aqueous solution.

Similar arguments show that the 'hydrogen ion' is $C_2H_5OH_2^+$ in ethanol solution, NH_4^+ in liquid ammonia, etc. This circumstance led to the introduction of a number of acid-base definitions explicitly involving the solvent: for example – 'An acid is a solute which gives rise to a cation characteristic of the solvent, and a base is a solute which gives rise to an anion characteristic of the solvent.' This kind of definition gives a logical account of behaviour in a particular solvent, and has been useful in stimulating work in unusual types of solvent. However, it does not give a clue to several of the general properties of acids and bases, such as catalysis and action on indicators, which often persist in solvents which do not give rise to cations or anions, or even in the complete absence of solvent. It is also inconvenient to use everyday words such as acid and base in a sense which varies with change of solvent.

A much more powerful and general definition of acids and bases was proposed almost simultaneously in 1923 by J. N. Brönsted in Denmark, and by T. M. Lowry in England. It reads: *An acid is a species having a tendency to lose a proton, and a base is a species having a tendency to add on a proton.* This can be expressed by the scheme:

$$A \rightleftharpoons B + H^+, \tag{1}$$

where A and B are termed a *conjugate (or corresponding) acid-base pair*. The definition places no restriction on the sign or magnitude of the charges on A and B, though, of course, A must always be more positive than B by one unit. It is important to realize that the symbol H^+ in this definition represents the bare proton, and not the 'hydrogen ion' of variable composition depending on the solvent (OH_3^+, $C_2H_5OH_2^+$, NH_4^+, etc.). The definition is thus independent of the solvent, but equation (1) represents a hypothetical scheme used for defining A and B, and not a reaction which can actually occur in solution.

The Brönsted-Lowry definition of an acid obviously includes the uncharged molecules known as acids in the classical dissociation theory, e.g. HCl, H_2SO_4, CH_3COOH, etc., and also anions like HSO_4^- and $COOH.COO^-$ which occur in acid salts. It further recognizes a class of cation acids which represent an extension of the older definition. For example, the ammonium ion is an acid in virtue

B

of the tendency represented by $NH_4^+ \rightleftharpoons NH_3 + H^+$, and we shall see later that the acid reaction of solutions of ammonium salts can be interpreted more simply in terms of the acid properties of the ammonium ion than by the older interpretation of hydrolysis. The definition of the ammonium ion as an acid will be valid in any solvent, and in particular in liquid ammonia, where it was regarded as the 'hydrogen ion' in older definitions. Similarly, the hydronium ion in water has now lost any claim to uniqueness, and is regarded merely as an important member of the class of cation acids.

The change in the definition of bases is more radical, since it appeals to the power of accepting a proton, and not to the production of OH^- or some similar ion. Uncharged bases include ammonia and the amines in virtue of schemes such as $RNH_2 + H^+ \rightleftharpoons RNH_3^+$, while the basic properties of the metallic hydroxides are attributed to the presence of the hydroxide ion, an important (but not unique) anion base. The class of anion bases includes the anions of all weak acids: for example, the alkaline reaction of a solution of sodium acetate is related to the basic nature of the acetate ion ($CH_3COO^- + H^+ \rightleftharpoons CH_3COOH$). It may be noted that according to the older definitions the acetate ion was regarded as the typical basic ion for solutions in glacial acetic acid, while the Brönsted–Lowry definition describes it as a base in any solvent. The restriction of basic properties to the anions of *weak* acids is only a matter of convenience: in principle the anion of any acid can be regarded as a base, but if the acid is a strong one (e.g. HCl, $HClO_4$) its anion has such a small tendency to accept a proton that its basic properties will not be detectable under ordinary circumstances. The same kind of practical restriction is of course applied in defining acids: any hydrogen compound can in principle be regarded as an acid, but in many of them (e.g. most hydrocarbons) the tendency to lose a proton is so small that they do not show acidic behaviour under ordinary conditions.

As already mentioned, the equation $A \rightleftharpoons B + H^+$ does not represent a process which can actually take place. However, if we combine two such equations we obtain:

$$A_1 + B_2 \rightleftharpoons A_2 + B_1, \tag{2}$$

and this is the general expression for all reactions involving acids and

Table 2

Examples of protolytic reactions

	A_1	+	B_2	⇌	A_2	+	B_1	Description
(a)	CH_3COOH		H_2O		OH_3^+		CH_3COO^-	Dissociation of acetic acid in water, or buffer action in acetic acid + acetate.
(b)	CH_3COOH		NH_3		NH_4^+		CH_3COO^-	Dissociation of acetic acid in liquid ammonia, or dissociation of ammonia in glacial acetic acid, or neutralization of acetic acid by ammonia (with or without solvent) or hydrolysis of ammonium acetate solution.
(c)	H_2O		CH_3COO^-		CH_3COOH		OH^-	Hydrolysis of sodium acetate solution, or neutralization of acetic acid by sodium hydroxide, or buffer action in acetic acid + acetate.
(d)	NH_4^+		H_2O		OH_3^+		NH_3	Hydrolysis of ammonium salts, or buffer action in ammonia + ammonium salt, or neutralization of ammonia by a strong acid.
(e)	H_2O		HPO_4^-		$H_2PO_4^-$		OH^-	Hydrolysis of secondary phosphates, or buffer action in a mixture of primary and secondary phosphates, or preparation of a secondary phosphate from a primary phosphate and sodium hydroxide.
(f)	H_2O		NH_3		NH_4^+		OH^-	Dissociation of ammonia in water, or buffer action in ammonia + ammonium salt.
(g)	$H_2PO_4^-$		H_2O		OH_3^+		HPO_4^-	Dissociation of primary phosphate in water, or buffer action in a mixture of primary and secondary phosphate, or preparation of a primary phosphate from secondary phosphate and a strong acid.

bases. Because of the relations $A_1 \rightleftharpoons B_1 + H^+$, $A_2 \rightleftharpoons B_2 + H^+$, equation (2) represents the *transfer of a proton* from A_1 to B_2 (or from A_2 to B_1), and reactions between acids and bases are therefore often called *protolytic reactions*. Table 2 gives some examples of the way in which various reactions of acids and bases, commonly described by a variety of names, can be expressed in the form of equation (2). Most of the examples refer to aqueous solutions, and it will be seen that the water molecule is acting as a base in reactions (*a*), (*d*), and (*g*), and as an acid in reactions (*c*), (*e*), and (*f*). Because of this dual function water is known as an *amphoteric* solvent.

Finally, it should be noted that the Brönsted–Lowry definition does not include some of the substances formerly classified as acid-base systems in special types of solvent. For example, it has been maintained that in liquid sulphur dioxide the typical ionization process is $2SO_2 \rightleftharpoons SO^{++} + SO_3^{=}$, the cation being characteristic of acids in this solvent and the anion characteristic of bases. For this reason thionyl chloride has been classed as an acid in liquid sulphur dioxide, and potassium sulphite as a base. This nomenclature would not be used in the Brönsted–Lowry system, since the ionization processes concerned do not involve the transfer of a proton. The question of extending the acid-base definition to systems not involving protons will be considered in Chapter 8 of this book.

GENERAL REFERENCES

P. Walden, *Salts, Acids and Bases* (Ithaca, N.Y., 1929).

N. Bjerrum, 'Salts, Acids and Bases', *Chem. Reviews*, 1935, **16**, 287.

R. P. Bell, 'The Use of the Terms Acid and Base', *Chem. Soc. Quarterly Reviews*, 1947, **1**, 113.

R. P. Bell, *The Proton in Chemistry* (Ithaca, N.Y., and London, 1959), Chapter II.

M. Eigen, *Pure Appl. Chem.*, 1963, **6**, 97.

CHAPTER 2
Acid-base Equilibria in Water

Reactions between acids and bases are frequently to some extent reversible, and the law of mass action must therefore apply to the equilibrium concentrations of the various reacting species. The expressions thus obtained appear under a variety of names (dissociation, hydrolysis, buffers, indicators, etc.), but the principles involved are exactly the same throughout, since all the equilibria concerned can be written in the form $A_1 + B_2 \rightleftharpoons A_2 + B_1$ (cf. equation (2)). If we apply the law of mass action in its simplest form to this equilibrium we obtain:

$$\frac{[A_2][B_1]}{[A_1][B_2]} = K(T), \tag{3}$$

where the constant $K(T)$ depends only on the temperature and the nature of the solvent (and not on the concentrations of the various reacting species). This simple expression is strictly valid only for extremely dilute solutions, since when it is derived theoretically (whether by kinetic or thermodynamic methods) it is necessary to assume that the species concerned obey ideal laws of some kind: for example, any forces which the solute particles exert on one another must have a negligible effect upon their properties. In equilibria involving only uncharged molecules this restriction is not an onerous one, since experiment shows that the ideal laws are obeyed with fair accuracy up to moderate concentrations. However, when ions are present the electrostatic forces between them have an appreciable effect on the properties of their solutions, and the ideal laws are no longer obeyed accurately even in fairly dilute solutions. The deviations from the ideal laws are usually expressed in terms of activities or activity coefficients. In the equilibrium, $A_1 + B_2 \rightleftharpoons A_2 + B_1$, at least two of the species concerned must be ions, and we must therefore expect that equation (3) will not be exactly valid over a

range of concentrations, and that the equilibrium will be disturbed
by the addition of other ions to the solution. In spite of these reserva-
tions we shall in this chapter deal only with the simple form of the
equilibrium expressions, deferring until Chapter 4 the consideration
of interionic attraction and ionic activities. The equations of the
present chapter will be approximately valid in solutions where the
total ionic concentration is low: moreover (as will be shown in
Chapter 4) equations of the same form will also hold at higher ionic
concentrations provided that the total ionic concentration is not
allowed to vary much in a given set of experiments.

The commonest use of equation (3) is for defining and measuring
the *strength of an acid*. If A_1 is a very much stronger acid that A_2 the
reaction $A_1+B_2 \rightarrow A_2+B_1$ will go almost completely from left to
right, and it is a natural extension of this idea to use the constant
$[A_2][B_1]/[A_1][B_2]$ as a measure of the relative strengths of A_1 and A_2.†

If we choose A_2—B_2 as a standard acid-base pair, the values of K
for different acids A_1 can be used as a measure of their strengths. In
principle any acid-base pair could be chosen as a standard, but it is
usually most convenient to refer acid-base strengths to the *solvent*;
for example, in water the acid-base pair OH_3^+—H_2O is taken as the
standard. The equilibrium used in defining acid strengths is therefore:

$$A+H_2O \rightleftharpoons B+OH_3^+, \qquad (4)$$

and the constant

$$\frac{[B][OH_3^+]}{[A][H_2O]} = K', \qquad (5)$$

gives the acid strength of A_1, that of the ion OH_3^+ being taken as
unity.

Equation (4) represents what is commonly described as the dissociation of the acid A in water, and the constant K' is closely related
to the *dissociation constant* of A as usually defined, differing only in
the inclusion of the term $[H_2O]$ in the denominator. This term repre-

---

† If the equilibrium $A \rightleftharpoons B+H^+$ had any real existence, the constant K in
equation (3) could be regarded as the ratio of the two constants $K_1 = [B_1][H^+]/$
$[A_1]$ and $K_2 = [B_2][H^+]/[A_2]$, but in view of what has been said about the free
proton, this is a somewhat artificial point of view.

sents the 'concentration' of water molecules in liquid water, and would have the value 55·5 moles per litre on the ordinary volume concentration scale. It is certainly not legitimate to use concentrations as high as this in a simple mass-law expression, even for uncharged molecules, and it would be difficult to determine the proper correction to apply. Fortunately, however, as long as we are dealing with reasonably dilute solutions in water at a given temperature, the value of $[H_2O]$ (and of any corrections associated with it) will be the same throughout. It can therefore be absorbed into the constant in equation (5) giving:

$$\frac{[B][OH_3^+]}{[A]} = K \tag{6}$$

as defining the strength of the acid A. For aqueous solutions there will be no ambiguity in using the abbreviation H^+ for OH_3^+ as long as it is realized that the hydrated proton is meant, and this abbreviation will be used extensively in the remainder of this book. If A is an uncharged molecule (e.g. an organic acid), B is the anion derived from it by the loss of a proton, and (6) is the ordinary expression for the dissociation constant. As a particular case A may be the water molecule when the appropriate dissociation constant is $[H^+][OH^-]/[H_2O]$. As before, the term $[H_2O]$ cannot vary in dilute aqueous systems at a fixed temperature, and it is therefore more usual to describe the dissociation of water in terms of the *ionic product* $K_w = [H^+][OH^-]$. This product has the value $1·0 \times 10^{-14}$ at 25°, but is strongly temperature dependent, being $1·0 \times 10^{-15}$ at 0°, and 7×10^{-13} at 100°. K_w is not directly comparable with the dissociation constants of other acids, and if we wish to specify the acid strength of the water molecule it is better to use the constant $K_w/[H_2O] = K_w/55·5 = 1·8 \times 10^{-16}$ at 25°, though the numerical value has no exact significance on account of the high value of $[H_2O]$ and the associated nature of water.

It is an obvious extension to specify the strength of an *anion acid* such as $H_2PO_4^-$ by a dissociation constant $[HPO_4^=][H^+]/[H_2PO_4^-]$, often referred to as the second dissociation constant of phosphoric acid. An exactly similar treatment applies to *cation acids*: for example, the acid strength of the ammonium ion is given by the expression $[NH_3][H^+]/[NH_4^+]$, which is completely analogous to a dissociation constant such as $[CH_3COO^-][H^+]/[CH_3COOH]$, only the charges being different. On this system the acid strength of the *hydronium ion* would be $[H_2O][OH_3^+]/[OH_3^+] = [H_2O] = 55·5$, remembering that the symbol H^+ is really an abbreviation for OH_3^+. Because of the difficulties connected with $[H_2O]$ mentioned above, not much weight can be attached to this value, but it shows that OH_3^+ must be considered as an extremely strong acid in aqueous solution.

The question of the *strength of bases* used to be treated separately from acids, but this is quite unnecessary, since any protolytic reaction involving an acid must also involve its conjugate base. Equations (3)–(6) therefore suffice to give us information about the strengths of bases as well as of acids. If the reaction $A_1 + B_2 \rightarrow A_2 + B_1$ goes almost completely from left to right, then B_2 is a much stronger base than B_1, and in general the order of strengths of a series of bases will be the inverse of the order of strengths of their conjugate (or corresponding) acids. It has therefore been proposed that the strength of a base should be represented by the reciprocal of the dissociation constant K of the conjugate acid, but this suggestion has not been generally adopted. It is simpler to retain the one dissociation constant K (cf. equation (6)) for defining the properties of the acid-base pair A—B, remembering that a greater value of K corresponds to a greater acid strength for A, but a smaller basic strength for B. In practice the values of K for different systems vary through many powers of ten, and it is often convenient to use instead the *dissociation exponent* pK, defined by:

$$pK = -\log_{10}K. \tag{7}$$

Because of the negative sign in (7), a larger pK corresponds to a weaker acid and a stronger base: this corresponds to the commonly used symbol pH $(= -\log_{10}[H^+])$, which has a greater value in less acid solutions. In particular, the pairs OH_3^+—H_2O and H_2O—OH^- have pK values of $-1·7$ and $+15·7$ respectively, showing that H_2O is a very weak base and OH^- a very strong one.

The older method of defining basic strength depended upon the concentration of hydroxide ions produced in an aqueous solution of the base: for example, the basic dissociation constant of ammonia was defined by†:

$$\frac{[NH_4^+][OH^-]}{[NH_3]} = K_b. \tag{8}$$

Since $[H^+][OH^-] = K_w$, we have $K_b = K_w/K$, where $K = [NH_3][H^+]/[NH_4^+]$ is the acidic constant of the ammonium ion. In the same way the 'conventional' basic dissociation constant of any base can be expressed in terms of the acidic constant of its corresponding acid, and in the remainder of this book we shall use only the latter type of constant.

The reaction $NH_4^+ \rightleftharpoons NH_3 + H^+$ (an abbreviation for $NH_4^+ + H_2O \rightleftharpoons NH_3 + OH_3^+$), used to define the strength of the ammonium ion, is identical with what is usually termed the *hydrolysis* of ammonium salts (cf. Table 2). The solution must of course also contain an equivalent concentration of anions, but if these have no appreciable acidic or basic properties (e.g. chloride, nitrate) they will take no part in the hydrolysis equilibrium. The constant K for NH_4^+ is in fact what is commonly termed the *hydrolysis constant* for an ammonium salt. The description of the equilibrium given here is simpler than the older picture of hydrolysis. The hydrolysis of the sodium salt of a weak acid can be similarly treated: for example, in a solution of sodium acetate the reaction is $OAc^- + H_2O \rightleftharpoons HOAc + OH^-$, the sodium ions playing no part, and the *hydrolysis constant* is $[HOAc][OH^-]/[OAc^-] = K_h = K_w/K$, where K is as usual the dissociation constant of acetic acid.

The types of acid-base equilibrium which we have so far considered can be given a common quantitative treatment. If a solution is made up containing a total concentration c of acetic acid,

† In this equation (and in similar ones in this book) the symbol $[NH_3]$ represents the total concentration of undissociated ammonia, irrespective of whether it is present as NH_3 or NH_4OH. The two species are related by the equilibrium $NH_3 + H_2O \rightleftharpoons NH_4OH$, and since water is present in large excess the ratio $[NH_3]/[NH_4OH]$ will be the same in all dilute solutions at the same temperature: hence the form of any equilibrium expressions will not be affected by taking into account partial hydration of the undissociated ammonia. It is in fact difficult to et reliable evidence as to the actual concentration of NH_4OH, and doubts have been expressed as to whether it exists at all: it will at most be a loose hydrogen-bonded complex between an ammonia and a water molecule.

ammonia, ammonium chloride, or sodium acetate, the concentrations of the various species at equilibrium can be represented in the following scheme:

$$
\left.\begin{array}{lll}
\text{(i)} & \text{HOAc} + \text{H}_2\text{O} \rightleftharpoons \text{OAc}^- + \text{OH}_3^+ \\
\text{(ii)} & \text{NH}_3 + \text{H}_2\text{O} \rightleftharpoons \text{NH}_4^+ + \text{OH}^- \\
\text{(iii)} & \text{NH}_4^+ + \text{H}_2\text{O} \rightleftharpoons \text{NH}_3 + \text{OH}_3^+ \\
\text{(iv)} & \text{OAc}^- + \text{H}_2\text{O} \rightleftharpoons \text{HOAc} + \text{OH}^- \\
& c(1-x) \qquad\qquad cx \qquad\quad cx
\end{array}\right\}, \qquad (9)
$$

where x is the degree of dissociation or hydrolysis. In each case the equilibrium is governed by the equation:

$$
\frac{cx^2}{1-x} = K'. \qquad (10)
$$

The separate expressions for K' for each reaction have already been indicated. Equation (10) shows that the degree of dissociation or hydrolysis increases with increasing dilution. The exact calculation of x in terms of c and K' involves the solution of a quadratic, but this is rarely necessary in practice. When x is much smaller than unity, (10) simplifies to $cx^2 = K'$: this frequently gives a sufficiently accurate answer, and will in any case give a rough value of x which can be inserted in the denominator of (10) to give a further approximation. Equation (10) is in any case not quite exact, since in putting both concentrations on the right-hand side of (9) equal to cx we are implicitly neglecting either $[\text{H}^+]$ or $[\text{OH}^-]$: however, since $[\text{H}^+][\text{OH}^-] = 10^{-14}$, one of these concentrations must be less than 10^{-7}, so that the approximation is usually sufficiently good.

In the above examples the solutions were made by dissolving either the acidic or the basic component of a conjugate pair (if necessary together with a neutral ion such as Na^+ or Cl^-). A more general type of solution is obtained by deliberately varying the proportions of acid and base: such a solution is termed a *buffer solution*. For example, a solution containing arbitrary concentrations of acetic acid and acetate ion can be prepared by mixing solutions of acetic acid and sodium acetate, or by partially neutralizing a solution of acetic acid with sodium hydroxide, or by adding less than one equivalent of hydrochloric acid to a solution of sodium acetate.

Similarly, an ammonia-ammonium ion buffer is made by mixing solutions of ammonia and ammonium chloride, or by partially neutralizing a solution of ammonia with hydrochloric acid, or by adding less than one equivalent of sodium hydroxide to a solution of ammonium chloride.

The hydrogen-ion concentration in a buffer solution is of course given by equation (6), which can be rewritten as:

$$[H^+] = K\frac{[A]}{[B]}, \text{ i.e. } pH = pK - \log_{10}\frac{[A]}{[B]}. \tag{11}$$

The ratio $[A]/[B]$ is termed the *buffer ratio*. It can often be calculated directly from the quantities used to make up the solution: for example, if 0.4 equivalents of sodium hydroxide are added to a solution containing one equivalent of acetic acid, the buffer ratio will be nearly $(1-0.4)/0.4 = 1.5$. This method of calculating is always sufficiently accurate for solutions not far from the neutral point ($pH = 7$), but if pH is smaller than 4 or greater than 10 the values of $[A]$ and $[B]$ in the solution will differ appreciably from the stoichiometric values $[A]^*$ and $[B]^*$. It is easily shown (best by using the conditions of electrical neutrality) that the correct equations are then:

$$\left.\begin{aligned} \text{for low pH, } \quad pH &= pK - \log_{10}\frac{[A]^* - [H^+]}{[B]^* + [H^+]} \\ \text{for high pH, } pH &= pK - \log_{10}\frac{[A]^* + [OH^-]}{[B]^* - [OH^-]} \end{aligned}\right\} . \tag{12}$$

As long as the corrections are small, $[H^+]$ and $[OH^-]$ on the right-hand side of (12) can be calculated with sufficient accuracy from (11). Failing this, equation (12) must be solved as a quadratic for $[H^+]$ or $[OH^-]$.

Buffer solutions are normally employed under conditions such that the corrections in (12) are not necessary, i.e. when $[A]$ and $[B]$ are both much greater than either $[H^+]$ or $[OH^-]$. (This contrasts with the position in dissociation or hydrolysis equilibria, where either the acidic or the basic component has a concentration comparable with $[H^+]$ or $[OH^-]$.) Buffer solutions are used for stabilizing the concentration of hydrogen or hydroxide ions in a solution: equation (11)

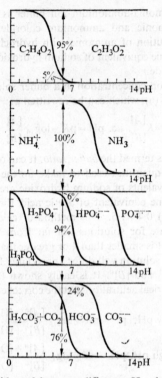

Fig. 1. State of acids and bases at different pH values for acetic acid, ammonia, phosphoric acid, and carbonic acid. (From N. Bjerrum: 'Inorganic Chemistry', 1936. *Heinemann*.)

shows that these concentrations depend only upon the *ratio* of $[A]$ to $[B]$, and are therefore unaffected by dilution with water, or with any solution not possessing acidic or basic properties.† Moreover, the pH of a buffer solution will be little affected by the addition of small quantities of strong acid or base, since these can interconvert

† It will be shown in Chapter 4 that this conclusion is only approximate, being modified considerably by a consideration of interionic forces.

only small quantities of A and B, thus producing little change in their ratio. This buffering action will be impaired if either $[A]$ or $[B]$ becomes too small: hence buffer ratios must not deviate too far from unity, and the effective buffering range of a given acid-base system is roughly from $pH = pK + 1$ to $pH = pK - 1$, corresponding to buffer ratios between $0 \cdot 1$ and 10. Figure 1, which is self-explanatory, shows the relation between pH and composition for a number of typical buffer systems.

A somewhat similar problem arises when considering the hydrolysis of a salt of a weak acid and a weak base, for example, the ammonium salt of a weak acid HX. We shall work out this example in detail to illustrate how the calculation can be simplified by suitable approximations. If c is the total concentration of the ammonium salt, and x and y the concentrations of ammonia and HX respectively formed by hydrolysis, then the following equations hold:

(i) $\dfrac{[H^+][NH_3]}{[NH_4^+]} = \dfrac{[H^+]x}{c - x} = K_1;$

(ii) $\dfrac{[H^+][X^-]}{[HX]} = \dfrac{[H^+](c - y)}{y} = K_2;$

(iii) $[H^+][OH^-] = K_w;$

(iv) $[H^+] + (c - x) = [OH^-] + (c - y)$ (electroneutrality condition).

We thus have four equations to determine the four unknowns, x, y, $[H^+]$ and $[OH^-]$. Writing h for $[H^+]$, and substituting from (i), (ii) and (iii) for x, y and $[OH^-]$ in (iv), we obtain the quartic equation

$$hc(K_1 K_2 - h^2) + (K_w - h^2)(K_1 + h)(K_2 + h) = 0,$$

which can be solved for h and hence for the other unknowns. A simpler solution results if we remember that on account of the small value of K_w at least one of the concentrations $[H^+]$ and $[OH^-]$ will be negligible compared with c. If we suppose that $[OH^-]$ can be neglected, (iv) becomes $x = y + h$, and the quartic equation reduces to a cubic,

$$c(K_1 K_2 - h^2) - h(K_1 + h)(K_2 + h) = 0.$$

A similar procedure is followed if $[H^+]$ is neglected instead of $[OH^-]$. A much greater simplification follows if both $[H^+]$ and $[OH^-]$ are negligible compared with c, as is often the case. Equation (iv) then becomes $x = y$, and (i) and (ii) give immediately:

$$\frac{x}{c-x} = \left(\frac{K_1}{K_2}\right)^{\frac{1}{2}}, \quad [H^+] = (K_1 K_2)^{\frac{1}{2}}.$$

In practice these simple equations would be examined first to determine which of the approximations is legitimate in any particular case. The last equation predicts that the degree of hydrolysis and the pH of the solution in question will be independent of its concentration, thus contrasting with the simpler type of hydrolysis shown in (9) and (10). A similar treatment can be applied when the acid and the base are not present in equivalent amounts, e.g. in the titration of a weak acid by a weak base.

In using a buffer solution an acid-base pair is added to the system in relatively high concentration, so as to control the concentrations of hydrogen and hydroxid ions, and the positions of any other acid-base equilibria in the system. In using *indicators* the position is reversed, since we add an acid-base pair in such small concentration that it has no appreciable effect on the state of the system, and then use observations of the relative concentrations of the added acid and base to draw conclusions about the pH of the system being investigated. The proportions of the indicator present in the acidic and basic forms are usually determined by means of the colour, and it is therefore necessary that the indicator should have a strong light absorption in at least one of its forms, and that the two forms should differ considerably in their absorption curves. These conditions are usually fulfilled by using rather complex organic molecules (e.g. triphenylmethane dyes) in which the change of colour on the addition or removal of a proton is associated with a drastic change in the bond structure of the molecule, for example, a change from a benzenoid to a quinonoid ring. However, contrary to some earlier views, these structural changes play no essential part in indicator equilibria, which are fully described by the interconversion of a single conjugate pair A—B. If ultra-violet light is used in place of visible, exactly the same behaviour can be observed with acid-base systems of much simpler structure.

The indicator equilibrium is of course governed by equation (11), and hence if the ratio $[A]/[B]$ is measured, the hydrogen-ion concentration of the solution can be determined provided that the indicator constant K is known. $[A]/[B]$ can be measured approximately by methods involving visual comparison, and accurately by photo-electric colorimeters, which can also be used in the ultra-violet. The constant K can be measured by any of the standard methods mentioned at the end of this chapter, or by colorimetric measurements in solutions of known pH, preferably buffer solutions. Since its concentration is very small, it does not matter whether the indicator is added in the acidic or the basic form: the choice depends upon which is more readily obtained in a pure state.

One of the main uses of indicators is to determine the *end-point of titrations*. When solutions of an acid and a base are gradually mixed, the pH at any stage can be calculated from equation (12) or by the method given for treating the hydrolysis of salts. Curves representing the course of typical titrations are shown in Fig. 2. The variation of $[A_i]/[B_i]$ (i referring to the indicator) is given by curves like those in Fig. 1, which show that there will be an almost complete change from A_i to B_i in the pH range $pK_i - 1$ to $pK_i + 1$. A sharp end-point will therefore be obtained if this range is included in the abrupt change of pH which takes place near the equivalence point in the titration. Reference to Fig. 2 shows that almost any indicator will serve for the titration HCl—NaOH, since the pH changes very sharply from 3·5 to 10·5. The same will be true for any other titration of a strong acid with a strong base, since the acidic and basic species in their solutions are OH_3^+ and OH^- respectively, the other ions (such as Cl^- and Na^+) playing no part in the titration. In the titration CH_3COOH—NaOH the pH changes rapidly only over the range 7·5–10·5, and an indicator with a pK_i in this range is needed (e.g. phenolphthalein, with $pK_i \sim 9$). On the other hand, in the titration HCl—NH_3 the pH change in the neighbourhood of the equivalence point is 3·5–6·5: phenolphthalein would therefore show no end-point in this titration, and a suitable indicator would be methyl orange ($pK_i \sim 4$). If both the acid and the base are too weak, as in the titration CH_3COOH—NH_3, the curve shows that there is no considerable change of pH in the neighbourhood of equivalence, so that no choice of indicator will make this titration an accurate one. However, if either the acid or the base is made ten times stronger, as in the titration of formic acid with ammonia, the change becomes sharp enough for a reasonably accurate titration provided that an indicator of exactly the right pK_i is chosen.

We shall end this chapter by enumerating briefly the *experimental methods* available for investigating acid-base equilibria in water, of which those given in equation (9) are typical. Details can be found in

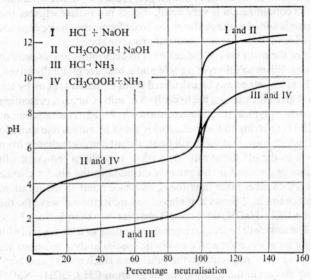

Fig. 2. Titration curves for N/10 solutions of strong and weak acids and bases.

a text-book on experimental electrochemistry. If ions appear on only one side of the equation, as in (9(i)) and (9(ii)), measurement of *electrical conductivity* will be suitable. This method is less frequently applicable to reactions such as (9(iii)) and (9(iv)), since the conductivity changes involved may be very small. Resort must then be had to methods which measure more specifically particular species concerned in the equilibria, in particular the ions OH_3^+ and OH^-. The most general method is to measure the potential of a *hydrogen electrode* (or some equivalent such as the quinhydrone or glass electrode) relative to a standard electrode: this gives a measure of the concentration of hydrogen ions (or hydroxide ions) in the system. We have just seen that the same object is attained by measuring the colour of suitable *indicator* in the solution: to attain reasonable accuracy it is clear that the pK of the indicator must not be far removed from the pH of the solution. A specific measure of the concentration of hydrogen or hydroxide ions can also be obtained by studying the velocity of certain *catalysed reactions*, which will be mentioned again in Chapter 6.

In particular cases use can be made of specific properties of other reactants. For example, if some of the species concerned are coloured the system will act as its own indicator: thus the state of dissociation of the nitrophenols can be investigated *colorimetrically*, since the anions are coloured, but the undissociated acid molecules are not. The concentration of an uncharged species can sometimes be estimated by measurements of *partition with another phase*. For example, the concentration of free aniline in an aqueous solution can be determined by measuring *partition coefficients* with an immiscible solvent such as benzene, which does not dissolve aniline in any other form: similarly, the proportion of ammonia present in the undissociated state in an aqueous solution can be derived from measurements of its *partial vapour pressure*. Each of these methods can be used in suitable circumstances for measuring the strength of acid-base systems.

GENERAL REFERENCES

R. P. Bell, *The Proton in Chemistry* (Ithaca, N.Y., and London, 1959), Chapter III.

G. Kortüm, *Treatise on Electrochemistry* (2nd edn.) (Elsevier, 1965), Chapter 10.

E. J. King, *Acid-Base Equilibria* (Pergamon, 1965), Chapters 1–6.

c

CHAPTER 3

Acids and Bases in Non-aqueous Solvents

The behaviour of acids and bases varies profoundly with the nature of the solvent, and the fact that most of the early work on acids and bases dealt with aqueous solutions helped to obscure the important part played by the solvent, especially in solvating the proton. It is convenient to classify solvents in relation to the properties of water, though it is in many respects an abnormal solvent. Water has a high dielectric constant ($\varepsilon = 78 \cdot 5$ at $25°$), and is amphoteric, possessing both acidic and basic properties. (The associated nature of water is not in itself important in the present context, except in so far as it modifies the dielectric constant or acid-base properties.) Most other solvents have dielectric constants considerably less than that of water, and this has a considerable effect in modifying the properties of electrolytes in general. However, the particular behaviour of acids and bases is still more influenced by the wide variations which can occur in the acidic or basic properties of the medium, and we shall classify solvents according to these properties. Only a few examples will be taken from the wide range of available data.

1. Amphoteric solvents resembling water. This class includes primarily the *alcohols*, which are chemically similar to water, but which have considerably lower dielectric constants (CH_3OH, 32, C_2H_5OH, 25, n-C_3H_7OH, 20). Most of the quantitative investigation of these solvents has been by conductivity measurements (especially by H. Goldschmidt, P. Walden, and (Sir) H. B. Hartley in the period 1900–30), but some work has also been done by electrometric and indicator measurements. Because of the lower dielectric constants the effect of interionic forces is much greater than it is in water, and the exact interpretation of the data more difficult (*see* Chapter 4). However, there exists a fair amount of quantitative information about acid-base equilibria in the alcohols, especially methanol and ethanol.

The effect of dielectric constant can to some extent be eliminated by comparing first reactions such as the following:

$$\left.\begin{array}{ll}(a) & NH_4^+ + H_2O \rightleftharpoons NH_3 + OH_3^+ \\ (b) & NH_4^+ + ROH \rightleftharpoons NH_3 + ROH_2^+\end{array}\right\} \tag{13}$$

Each of these equilibria involves no change in the number of ions, and a change of dielectric constant alone will have little effect. The equilibrium constant for (13b) is obtained by combining data (obtained from conductivity measurements) on the dissociation $NH_3 + ROH \rightleftharpoons NH_4^+ + OR^-$ with the value for the ionic product of the solvent, $[ROH_2^+][OR^-] = K_s$. The value of K_s for the alcohols is so low (10^{-17} for CH_3OH and 10^{-19} for EtOH) that it is difficult to obtain the conductivity of the solvent, which is too sensitive to impurities, but it can be derived from measurements of electromotive force in the same way as K_w in water. It is found, for example, that in ethanol the equilibrium constant for (13b) is about six times as great as the corresponding (13a) for water: this can be expressed by saying that ethanol as a solvent is about six times as strong a base as water, when the effect of dielectric constant is eliminated.† Similar results are obtained with various amines in place of ammonia, and methanol instead of ethanol.

In the same way, the difference in acidic strength between water and the alcohols can best be studied apart from dielectric effects by equilibria such as:

$$\left.\begin{array}{ll}(a) & CH_3COO^- + H_2O \rightleftharpoons CH_3COOH + OH^- \\ (b) & CH_3COO^- + ROH \rightleftharpoons CH_3COOH + OR^-\end{array}\right\} \tag{14}$$

which can be derived by combining dissociation constant measurements with K_w or K_s. For carboxylic acids in ethanol the constant (14b) differs little from (14a), indicating that water and alcohol do not differ widely in acidic nature.

Much larger differences appear between water and the alcohols

† It should be noted that this comparison of basic properties refers to water and ethanol in the liquid state, where association will have a marked effect. The basic strengths of the H_2O and C_2H_5OH molecules are more properly investigated by studying their dilute solutions in another solvent.

when differences in dielectric constant play a part, as in equilibria of the following types:

$$(a) \quad R.NH_2 \ + H_2O \ \rightleftharpoons R.NH_3^+ + OH^-$$
$$(b) \quad R.NH_2 \ + EtOH \rightleftharpoons R.NH_3^+ + OEt^-$$
$$(c) \quad R.COOH + H_2O \ \rightleftharpoons R.COO^- + OH_3^+$$
$$(d) \quad R.COOH + EtOH \rightleftharpoons R.COO^- + EtOH_2^+ \tag{15}$$

In each of these, two ions are produced from none, a process which will be facilitated by a high dielectric constant. Correspondingly it is found that the basic dissociation constants of ammonia and the amines in ethanol are smaller by a factor of about 10^4 than in water, the corresponding factor for the acidity of carboxylic acids being about 10^5. It is also interesting to find that some acids which appear to be completely dissociated in water have quite small dissociation constants in the alcohols: for example, nitric acid has $K = 2\cdot5 \times 10^{-4}$ in ethanol. The fact that the ionic product (or autoprotolysis constant) of ethanol is about 10^5 times as small as that of water can also be attributed mainly to the difference in dielectric constant. A factor of 10^4–10^5 can in fact be accounted for reasonably on a purely electrostatic basis. The electrical free energy of a pair of separated ions of charges $+e$ and $-e$ and radius r in a medium of dielectric constant ε is given by elementary electrostatics as $e^2/\varepsilon r$, and the free energy difference of a pair of ions in two media of differing dielectric constants is thus:

$$\Delta G = \frac{e^2}{r}\left(\frac{1}{\varepsilon_1} - \frac{1}{\varepsilon_2}\right) \tag{16}$$

commonly known as the *Born equation*. If we apply this to water and ethanol ($\varepsilon_1 = 78$, $\varepsilon_2 = 25$) and give r the reasonable value 2×10^{-8} we find $\Delta G = -4\cdot3 \times 10^{-13}$ ergs per molecule, i.e. -6200 calories per gram-molecule. The corresponding ratio between equilibrium constants in water and ethanol is then $e^{-\Delta G/RT} = 3 \times 10^4$, which is of the observed order of magnitude. The separation of the observed effects into a dielectric effect and a chemical one has no exact quantitative significance, but it certainly seems justifiable to attribute a large part of the difference between water and the alcohols to their difference in dielectric constant, especially in view of their close chemical similarity.

2. Basic solvents. The only well-investigated solvents which are much more basic than water are *ammonia* and the *amines*. Liquid ammonia (b.p. $-33°C.$, $\varepsilon = 22$) has been extensively studied, especially by E. C. Franklin and C. A. Kraus, though not much of the data give useful information about the behaviour of acids and bases. Matters are often complicated by chemical reaction, and the solutions studied are mostly too concentrated for the quantitative interpretation of conductivity data, this being the experimental method most frequently employed. However, it is clear that ammonia is a much more strongly basic solvent than water, and that many acids which are weak in water (e.g. acetic acid) react almost completely according to the scheme $HA + NH_3 \rightarrow A^- + NH_4^+$, where the NH_4^+ ion is of course the 'hydrogen ion' in ammonia as solvent. For example $0 \cdot 1$ N solutions of acetic, benzoic, formic, thiocyanic, nitric, hydrochloric, hydrobromic, hydriodic and perchloric acids all have almost the same catalytic power in liquid ammonia, although their strengths differ widely in water. This catalytic power is clearly due to the ammonium ion, to which all these acids are converted almost completely, thus behaving as strong acids.[†] The range of acids which can exist as such in liquid ammonia is therefore a very limited one, and this phenomenon has been termed by Hantzsch the *levelling effect* of the solvent. It should be noted that the classification of a number of different acids as 'strong' in a given solvent does not mean that they have the same dissociation constants. For example, if three different acids are dissociated to the extent of 99%, 99·9% and 99·99% in N/10 solution, their dissociation constants will be 10^{+1}, 10^{+2}, and 10^{+3} respectively, but their solutions will be experimentally indistinguishable from the acid-base point of view since the solvated hydrogen ion will be effectively the only acid species present. The same

† It is necessary to distinguish here between complete *ionization* (or, more correctly, protolysis) and complete *dissociation*. The conductivities of the above solutions are not all the same: this is because the ions formed are not completely free, but are partly associated together by electrostatic forces. This behaviour is common in electrolyte solutions of low dielectric constant and moderate concentration, and will be dealt with further in the next chapter.

levelling effect occurs in water, though it is not so widespread because of the lower basic strength of this solvent: thus reasonably dilute aqueous solutions of HCl, HBr, HI, HNO_3, $HClO_4$ and the sulphonic acids are indistinguishable in their acid properties, though we shall see later that the choice of more suitable conditions reveals large differences between the strengths of these acids.

Liquid ammonia also has some acidic properties, as is shown by the existence of metallic amides, MNH_2, which are strong electrolytes in ammonia giving rise to the anion NH_2^-. However, its acidic properties are very weak, and little is known about its reaction with bases. The autoprotolysis reaction $2NH_3 \rightleftharpoons NH_4^+ + NH_2^-$ takes place only to a very small extent, $[NH_4^+][NH_2^-]$ having the value 10^{-33} at $-50°$.

There are a number of solvents which have basic properties similar to those of water, but which are devoid of acid properties: these include many organic oxygen compounds such as ethers, ketones, and esters. Their dielectric constants are low and little is known about their electrochemical behaviour, but the extent to which they react with acids can be estimated by studies of absorption of light (usually ultra-violet) with or without the addition of indicators. For example, E. A. Braude (1947–50) concluded that the basic strength of four solvents containing oxygen decreases in the order $H_2O >$ dioxane $>$ ethanol $>$ acetone, though the differences amount to less than a power of ten. The marked differences in the electrochemical behaviour of acids in these solvents is largely due to the variation of dielectric constant (H_2O, 78; dioxane, 2·2; ethanol, 25; acetone, 19).

3. Acidic solvents. Much work has been done in solvents which are considerably more acidic than water, notably in anhydrous *acetic acid* which has been studied especially by J. B. Conant and N. F. Hall (1927–40) and by I. M. Kolthoff and his collaborators (1934 onwards). Its dielectric constant is low (6·3), so that ionized compounds will be only partly dissociated into free ions, and conductivity measurements are difficult to interpret. Most of the information about acid-base equilibria in this solvent derives from measurements of absorption of light (visible or ultra-violet) or measurements of electromotive force. The latter are carried out by using a platinum electrode in contact with solid tetrachloroquinone and tetrachloro-hydroquinone (chloranil) which serves as a substitute for the hydrogen electrode as does the quinhydrone electrode in water. If acetic acid is used as a solvent for uncharged bases (e.g. amines) its strongly acid properties become apparent in a pronounced *levelling effect*. All bases which in water are stronger than aniline give almost identical titration curves with a strong acid, and must be assumed to react completely with the solvent according to the scheme $B + CH_3COOH \rightarrow BH^+ + CH_3COO^-$. For example, the aliphatic amines and the alkylanilines, all of which are weak bases in water, behave as strong bases in acetic acid. (It is interesting to note than there are very few uncharged bases which are 'strong' in this sense in water, the only common examples being the amidines, $R.C\diagdown\diagup\begin{smallmatrix}NH\\NH_2\end{smallmatrix}$.)

Much weaker bases, for example the nitranilines, react considerably with glacial acetic acid, and many substances which are extremely weak bases in water (e.g. urea, the oximes, and triphenylcarbinol) show measurable basic properties in acetic acid.

The ionic product of anhydrous acetic acid is not known with any certainty, but it is certainly less than 10^{-13}: since the acidic strength of acetic acid is much greater than that of water, its basic strength is much less. Correspondingly, only the strongest acids react apprecia-bly with acetic acid according to the scheme $HA + CH_3COOH \rightarrow CH_3COOH_2^+ + A^-$, and the levelling effect observed in aqueous solution is absent. The 'strong' acids $HClO_4$, HBr, H_2SO_4, HCl and HNO_3 form a series of decreasing strength, as revealed by their effect on the conductivity of the solvent and the potential recorded by a chloranil electrode. They also give different potentiometric titration curves if titrated with the same base. Because of the low dielectric constant of the solvent it is difficult to give any absolute values for ionization constants, but the following relative values have been estimated on the basis of conductivity measurements, the value for nitric acid being taken arbitrarily as unity: HNO_3, 1; HCl, 9; H_2SO_4, 30; HBr, 160; $HClO_4$, 400. As already mentioned, the dis-sociation constants of these acids in water may easily vary by similar factors, but they are all so great that they cannot be measured directly. Solutions of such acids in acetic acid have much stronger acidic properties than any aqueous solutions, and they can be used for titrating very weak bases such as oximes and amides, which cannot be estimated by titration in water. A solution of perchloric acid is the most suitable reagent, and the end-point can be detected either potentiometrically or by a suitable indicator.

Similar investigations have been carried out with anhydrous *formic acid* as solvent, especially by L. P. Hammett (1930–5). This is considerably more acidic than acetic acid, but otherwise its behaviour towards bases is similar. It has the advantage of a much higher dielectric constant ($\varepsilon = 62$), but its basic strength is fairly high, as is shown by the high autoprotolysis constant $[HCOOH_2{}^+]$ $[HCOO^-] = 10^{-6}$. For this reason its levelling effect on dissolved acids is not much less than that of water: thus $HClO_4$, H_2SO_4 and benzenesulphonic acid are all strong, though HCl is incompletely ionized. (The high dielectric constant is presumably due to association through hydrogen bonding, which implies that the molecule has a strong tendency both to donate and to accept a proton, so that there is an intimate connection between the high dielectric constant and the extensive autoprotolysis.) Since very strong acids such as $HClO_4$ are 'levelled down' by conversion to the ion $HCOOH_2{}^+$, formic acid is less suitable than acetic acid as a solvent for investigating or titrating very weak bases.

The most acid solvent which has been systematically investigated is *sulphuric acid*: the pioneer work is due to A. Hantzsch (1907–30), while more accurate measurements have been made by L. P. Hammett (1933–7), by R. J. Gillespie, and by P. A. H. Wyatt (1950 to present day). Although the basic strength of sulphuric acid is very low, on account of its extremely high acidity it has a high degree of autoprotolysis, with $[H_3SO_4{}^+][HSO_4{}^-] = 1\cdot7 \times 10^{-4}$, while the position is further complicated by another type of dissociation, $2H_2SO_4 \rightleftharpoons OH_3{}^+ + HS_2O_7{}^-$, with an equilibrium constant $[OH_3{}^+][HS_2O_7{}^-] = 7 \times 10^{-5}$. For this reason it is difficult to study the behaviour of solutes by means of conductivity or e.m.f. measurements, and most of the information has been derived from cryoscopic measurements. The freezing point of pure sulphuric acid is a convenient one

(10·36°C.), and it is possible to apply corrections for the self-ionization of the solvent. Just as for formic acid the high self-ionization is accompanied by a high dielectric constant, and the low volatility and viscous nature of sulphuric acid are also attributable to strong hydrogen bonding. It was at one time supposed that the dielectric constant of sulphuric acid was vastly greater than that of any other solvent, but recent measurements by J. C. D. Brand, J. C. James and A. Rutherford (1953) and by R. J. Gillespie and R. H. Cole (1956) have given values of 110 at 20° and 101 at 25°. Salts are therefore completely dissociated in this solvent, and corrections for interionic attraction are small.

Sulphuric acid is such a strongly acid medium that almost all compounds containing oxygen or nitrogen will accept a proton from it to some degree, thus behaving as bases. For example, not only amines, but also ethers, ketones, and esters give a twofold freezing-point depression, corresponding to complete reaction according to the scheme $>O + H_2SO_4 \rightarrow >^+OH + HSO_4^-$. Many substances normally regarded as acids exhibit basic properties in sulphuric acid: thus most carboxylic acids are strong bases, forming the ion $R.COOH_2^+$, though this reaction is incomplete for strong acids such as di- and trichloroacetic acids. Aliphatic and aromatic nitro-compounds, sulphones and sulphonic acids also behave as bases, though as weak ones. There are in fact very few compounds which will behave as simple non-ionized solutes in sulphuric acid, among them being sulphuryl chloride, chlorosulphonic acid, trinitrobenzene, trinitrotoluene and picric acid. Water ionizes almost completely as a base: it is surprising that this reaction is not quite complete, and it seems probable that the water molecule is stabilized by being incorporated in the highly associated structure of liquid sulphuric acid.

Very few substances exhibit acidic properties in sulphuric acid: indeed, as we have already seen, many typical acids behave as bases. Perchloric acid, often regarded as the strongest acid known, has a dissociation constant of 10^{-4}, and the only acid known to be stronger is $H_2S_2O_7$, for which $[H_3SO_4^+][HS_2O_7^-]/[H_2S_2O_7] = 2 \times 10^{-2}$. As might be expected, many substances when dissolved in sulphuric acid undergo reactions which are more complicated than simple proton-transfers: for example,

$$C_2H_5OH + 2H_2SO_4 \rightarrow C_2H_5.SO_4H + OH_3^+ + HSO_4^-;$$
$$(C_6H_5)_3C.OH + 2H_2SO_4 \rightarrow (C_6H_5)_3C^+ + OH_3^+ + 2HSO_4^-;$$
$$2.4.6.Me_3.C_6H_2.COOH + 2H_2SO_4 \rightarrow$$
$$2.4.6.Me_3.C_6H_2.CO^+ + OH_3^+ + 2HSO_4^-;$$
$$HNO_3 + 2H_2SO_4 \rightarrow NO_2^+ + OH_3^+ + 2HSO_4^-.$$

The carbonium ions formed in the above reactions are often important because of their chemical reactivity, and the nitronium ion NO_2^+ is frequently the active agent in the nitration of organic compounds: however, these questions are outside the scope of the present book.

Another strongly acid solvent of high dielectric constant is *hydrogen fluoride* ($\varepsilon = 84$ at $0°$) which has been extensively studied initially by K. Fredenhagen (1930–40), and later by M. Kilpatrick (1957 onwards) and others. The conductivity of the pure solvent is low, and this fact, coupled with the high dielectric constant, makes conductivity measurements the most suitable method of investigation. It is strange that hydrogen fluoride should be so strongly acidic as a solvent, since its dissociation constant in aqueous solution is only 7×10^{-4}, about the same as formic acid. The high acidity of the pure solvent is presumably related to its high degree of association by hydrogen bonding, which is also responsible for the high dielectric constant and low volatility of hydrogen fluoride. Its behaviour towards bases resembles that of sulphuric acid, though it is less extreme. A wide range of nitrogen and oxygen compounds behave as bases, though alcohols and phenols are weak, and trichloroacetic acid and acetyl fluoride behave as non-electrolytes. Few acids show any measurable dissociation in hydrogen fluoride, exceptions being $HClO_4$, HIO_4 and H_2SO_4, though in the latter case the reaction $H_2SO_4 + 3HF \rightarrow HSO_3F + OH_3^+ + HF_2^-$ also occurs. Many acids are both weak and insoluble: for example, if a metallic chloride or bromide is added to hydrogen fluoride there is an almost quantitative evolution of gaseous HCl or HBr.

4. Aprotic solvents. This term was originally used to describe solvents which have no tendency either to lose or to gain a proton, notably the *hydrocarbons* and their halogen derivatives. Relatively little work has been done in this type of solvent, but data exist for benzene, chlorobenzene, and chloroform. Because of the inert nature of the solvent it is clear that no dissociation or other reaction can take place when a single acid or base is dissolved: in particular, the levelling effect observed in other types of solvent is entirely absent. Observable acid-base reactions can occur only if both an acid and a base are added to the solvent, when the ordinary type of reaction $A_1 + B_2 \rightleftharpoons A_2 + B_1$ can take place. The occurrence of typical acid-base reactions (neutralization, indicator action) in aprotic solvents was one of the facts which led to the adoption of an acid-base definition independent of the solvent, as described in Chapter 1.

The non-participation of the solvent is in principle a simplification, but in practice various difficulties arise in the investigation of acids and bases in hydrocarbons and similar solvents. Their dielectric constant is low (usually 2–6) so that any ions formed will be considerably associated, and conductivity measurements yield little information. This association complicates the nature of the equilibria involved, and also extends to uncharged molecules: for example, carboxylic acids exist largely as dimers, and the large organic molecules commonly employed as indicators are often associated in such a way that Beer's law is not obeyed. The same factors lead to the low solubility of many substances in hydrocarbon solvents. On account of these difficulties our knowledge of acid-base equilibria in these solvents is mostly semi-quantitative in nature, though useful information has been obtained both by indicator studies and by e.m.f. measurements.

The term *aprotic* has been used recently in a wider sense to denote solvents which are unable to lose a proton, though they may be able to accept one: the contrast is with *protic* solvents, which are able to donate a proton, either in a complete transfer to a basic species, or to form a hydrogen bond, and *amphiprotic* solvents, which can act both as donors and acceptors of protons. (Most of the solvents so far considered in this chapter are amphiprotic.) Much attention has recently been paid to *dipolar aprotic solvents* (D. J. Cram,

A. J. Parker, M. C. Whiting, 1955 onwards), which frequently have high dielectric constants: examples are dimethylformamide ($\varepsilon=38$), dimethyl sulphoxide ($\varepsilon=49$) and nitrobenzene ($\varepsilon=35$). None of these solvents possess any appreciable proton-donating properties, and several of them are also extremely weak bases. Because of their relatively high dielectric constants ion-pair formation can often be neglected, though acid-base equilibria are often complicated by the formation of complexes such as HA_2^- between an acid and its anion. The dipolar aprotic solvents resemble the hydrocarbons in possessing a powerful differentiating effect upon the properties of acids and bases, and are frequently more useful because of their greater solvent power for a variety of electrolytes and non-electrolytes. In particular, basic anions are poorly solvated in these solvents, and thus behave as much stronger bases: for example, it has been estimated that sodium methoxide dissolved in dimethyl sulphoxide is 10^9 times as strong a base as it is in methanol, and the addition of even moderate proportions of dimethyl sulphoxide to aqueous sodium hydroxide greatly increases its activity towards acidic indicators or as a basic catalyst.

5. The strengths of acids in different solvents.

The conventional method of defining the strength of an acid by its dissociation constant implies a comparison with the solvent, and we have seen in the earlier part of this chapter that the extent of a reaction such as $HA+S \rightleftharpoons A^-+SH^+$ (where S is the solvent) depends in a complex manner on the chemical properties of the solvent and on the dielectric constant. We shall therefore not expect to find any simple regularities in the conventional dissociation constants of a given acid in different solvents, even when these are experimentally accessible. There has been much discussion as to whether it is possible to compare the 'absolute' acid strengths of a substance in two different solvents, but this has not been very fruitful, being bound up with the problem of defining the difference of electrical potential between different phases. Of greater practical interest is the question of how the *relative strength* of two acids depends upon the medium. This can be measured by a direct study of the equilibrium $A_1+B_2 \rightleftharpoons A_2+B_1$, or the two acid-base pairs may be successively brought into equilibrium with a standard pair $A°—B°$, and the ratio of the equilibrium constants taken. If $A°—B°$ is the solvent pair $SH^+—S$, this amounts to comparing the dissociation constants of the two acids, while if $A°—B°$ is an indicator system the comparison can be made directly from the indicator measurements. The value obtained of the relative strength of a particular pair of acids in a given solvent should of course be independent of the method used for measuring it. When we review the data for different solvents it is found that *the relative strengths of any two acids are approximately independent of the nature of the solvent*. This generalization is only true if the acids compared are of the same charge-type, but under these conditions it holds to within about a power of ten, provided that the acids studied do not differ too widely in chemical type. It has been established by the study of solvents of many different types, including methanol, ethanol, *n*-butanol, *m*-cresol, formamide, acetonitrile, acetic acid, chloroform, chlorobenzene, and benzene. An example of the kind of data obtained is given in Fig. 3 for the strengths of a number of acids relative to benzoic acid both in water and in *n*-butanol, the latter data coming from electrometric measurements by L. A. Wooten and L. P. Hammett (1935). If the relative strengths were exactly the same in the two solvents all the points would lie on the straight line of unit slope.

This parallelism between acid-strengths in different solvents makes it possible to estimate roughly the pK of many acid-base pairs in water even when their pK is either too high or too low for direct measurement: for example, by using data in non-aqueous solvents we can estimate that HCl has a pK of about -7 (i.e. a dissociation constant of 10^{+7}), while triphenylcarbinol has pK about $+19$. These estimated pK values in water can be used to give a general survey of the acid-base properties of the different types of solvent already described. In a given solvent (possessing both acidic and basic properties) any acid-base pair with a pK in water greater than a certain value pK' will be present almost entirely in the acidic form; similarly, if pK is smaller than another value pK'' it will be present almost entirely in the basic form. The values pK' and pK'' are characteristic of the solvent, and the range between them represents the range of acid-base pairs whose equilibria can be conveniently

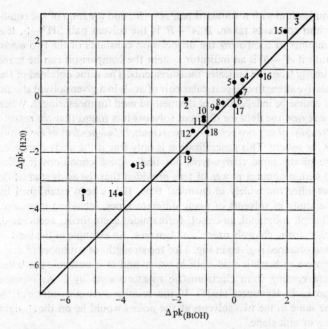

Fig. 3. Relative strengths of acids in water and in *n*-butanol referred to benzoic acid: 1 picric; 2, 2,4-dinitrophenol; 3, *p*-nitrophenol; 4, butyric; 5, acetic; 6, *m*-toluic; 7, *p*-toluic; 8, *p*-chlorobenzoic; 9, *m*-chlorobenzoic; 10, *m*-nitrobenzoic; 11, *p*-nitrobenzoic; 12, salicylic; 13, maleic (1); 14, trichloroacetic; 15, maleic (2); 16, trimethylacetic; 17, *o*-toluic; 18, *o*-chlorobenzoic; 19, *o*-nitrobenzoic. (From L. P. Hammett: 'Physical Organic Chemistry', *McGraw-Hill*.)

investigated in the solvent in question. If $pK < pK''$ the acid will be 'levelled' by conversion to the solvent cation, while if $pK > pK'$ the base will be similarly 'levelled' by conversion to the solvent anion. Figure 4 shows the approximate ranges for some common solvents: the figures are only rough, since they do not take into account variations in structure and charge, and in any case reliable data are lacking for several of the solvents. For the solvent ether no limit of pK is given on the upper side, because this solvent has no detectable acidic properties. Similarly, the scale for benzene extends indefinitely in both directions, since it behaves neither as an acid nor as a base. No solvent has been studied which possesses acidic properties but not basic ones.

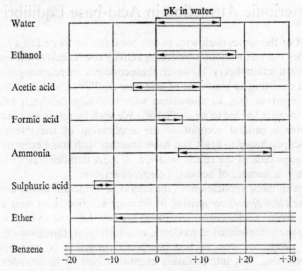

Fig. 4. Range of existence of acids and bases in different solvents.

GENERAL REFERENCES

G. Jander, *Die Chemie in wasserähnlichen Lösungsmitteln* (Berlin, 1949).

L. F. Audrieth and J. Kleinberg, *Non-Aqueous Solvents* (New York, 1953).

G. J. Janz and S. S. Danyluk, 'Conductances of Hydrogen Halides in Anhydrous Polar Organic Solvents', *Chem. Reviews*, 1960, **60**, 209.

A. J. Parker, 'Solvation in Dipolar Aprotic Solvents', *Chem. Soc. Quarterly Reviews*, 1962, **16**, 163.

E. J. King, *Acid-Base Equilibria*, Chapter 11 (Oxford, 1965).

J. J. Lagowski (Ed.), *The Chemistry of Non-Aqueous Solvents* (New York and London, 1966).

R. P. Bell, *The Proton in Chemistry* (Ithaca, N.Y., and London, 1959), Chapter IV.

CHAPTER 4

Interionic Attraction in Acid-base Equilibria

Most of the topics dealt with in this book can be treated in a satis-factory manner without introducing activity coefficients or the inter-ionic attraction theory. However, these concepts are necessary in any exact quantitative treatment of acid-base equilibria, and there are a few instances (e.g. in connection with acid-base catalysis) where their neglect has led to serious errors. We shall therefore give in this chapter a general account of the application of the interionic attraction theory to acids and bases together with some estimate of the magnitude of the effects involved. A more detailed treatment is given in a number of books on electrochemistry.

The simplest applications of the interionic attraction theory are to *strong electrolytes*, comprising in aqueous solution most salts and a few strong acids. These electrolytes are now believed to be almost completely dissociated at moderate concentrations, though many of their properties were at first interpreted in terms of incomplete dissociation. This interpretation eventually had to be abandoned because the study of various properties of the same solution (con-ductivity, thermodynamic properties, and optical properties) led to inconsistent values for the degree of dissociation, and because none of the values satisfied the simple law of mass action. Some of the properties (notably optical ones) indicated degrees of dissociation very close to 100%, and it was evidence of this kind which led to the assumption of complete dissociation. The apparently incomplete dissociation shown by other properties of the solution was now attributed to the electrical forces between the ions. Calculation shows that these forces are very great, and their effect will be greater in more concentrated solutions, since the ions are on the average closer together. These interionic forces can be made to account for a de-crease in ionic mobility with increasing ionic concentration, and also for apparently incomplete dissociation in thermodynamic pheno-mena such as osmotic pressure, freezing-point depression, solubilities, etc. The simple laws for all these last phenomena are related to Henry's law, which will be valid only if we can neglect any forces between the solute molecules which is not the case for ionic solutions. As might be expected the effect of the interionic forces is greatest for multiply charged ions and for solvents of low dielectric constant.

If positive and negative ions were distributed in a random manner in solution, the net effect of the interionic forces would be zero, since repulsions would be just as frequent as attractions. The actual distribution is not a random one, but represents a compromise between the Coulomb potential energy, which tends to bring positive and negative ions together, and the thermal motion of the ions, which tends to produce randomness. The resulting average distribution is called the *ionic atmosphere*, a term which emphasizes its fluid and statistical nature. If we consider a small element of volume in the neighbourhood of a particular ion, the probability that it contains an ion of opposite charge is somewhat greater than the probability that it contains an ion of like charge. These two probabilities will become more unequal as the solution becomes more concentrated, while in a very dilute solution they will be equal, corresponding to a random distribution. The expansion of an ionic atmosphere by dilution therefore involves the separation of some ions of opposite charge, and the work involved in this separation can be related directly to the thermodynamic properties of strong electrolyte solutions.

The Debye-Hückel theory of electrolytes consists of a quantitative treatment of this picture, and we shall give some of its results here. It is possible to derive equations both for the conductivity and for the thermodynamic properties of electrolyte solutions, of which the latter are the more important for our purpose. The effect of the interionic atmosphere upon these properties can be expressed in various ways, of which the most convenient is the *activity coefficient*. This can be regarded as a correction factor by which ionic concentrations must be multiplied before they can be inserted in various classical thermodynamic expressions, for example, the law of mass action or the expression for the e.m.f. of a concentration cell. All ionic activity coefficients become unity in sufficiently dilute solution, and they are usually less than unity in solutions of finite but not too high

D

ionic concentration. The interionic attraction theory gives a simple theoretical expression for ionic activity coefficients in dilute solutions. It is convenient first to define a quantity known as the *ionic strength*, by the equation:

$$I = \tfrac{1}{2}\Sigma m_i z_i^2, \tag{17}$$

where m_i and z_i are respectively the molar concentration and valency of a particular type of ion, and the summation is taken over all the ions present in solution. The definition of ionic strength applies equally to solutions containing a single electrolyte and to mixtures, and in the particular case of a single 1-1-valent electrolyte it reduces to m, the concentration of the electrolyte. The activity coefficient of an ion of valency z is given by:

$$-\log_{10} f_z = z^2 A I^{\frac{1}{2}} / (1 + B I^{\frac{1}{2}}), \tag{18}$$

where A and B are constants. A is given by the theory in terms of the temperature and the dielectric constant of the solvent and has the same value for any electrolyte, while B contains the mean ionic radius and will therefore vary to some extent from case to case. For aqueous solutions near room temperature A has a value of 0·50, while the average value of B is close to unity, so that a useful approximate form of (18) is:

$$-\log_{10} f_z = 0.50 z^2 I^{\frac{1}{2}} / (1 + I^{\frac{1}{2}}). \tag{19}$$

Equation (18) has been confirmed experimentally for univalent ions at concentrations up to about $I = 0.1$, while its range of validity is somewhat less for ions of higher valency. Equation (19) can be used to estimate activity coefficients even at higher concentrations, though the specific differences between different electrolytes will then become important. The distinguishing feature of these equations is the occurrence of the *square root* of the concentration, and it is worth emphasizing that no assumptions about incomplete dissociation can ever lead to an expression containing the concentration raised to a power less than one. The occurrence of the square root in the interionic theory is the result of two circumstances, firstly the long-range nature of the forces between the ions, and secondly the presence of two different kinds of particles, which attract when of opposite charge and repel when of like charge. The experimental verification of the square-root law thus provides compelling evidence that the behaviour of strong electrolytes is due to interionic forces rather than to incomplete dissociation.

The application of the interionic theory to strong electrolytes is not of primary importance in acid-base equilibria, except in so far as it shows the illusory nature of the degrees of dissociation derived by classical methods for strong acids such as HCl and HNO_3. The most important application is to *weak electrolytes*. As long as a solution contains only a weak electrolyte the ionic concentrations are low and the ideal laws will be approximately obeyed, as is shown by the wide success of the classical theory of electrolytic dissociation. The position is otherwise, however, if the salt concentration is increased either by making a buffer mixture, or by adding a neutral salt to a solution of a weak electrolyte. Anomalies in systems of this kind have been long known, under the name of *salt effects*. The magnitude of the effect is shown by Table 3, which gives the observed dissociation constant of acetic acid in solutions of potassium chloride and strontium chloride, derived from electromotive force measurements.

The effect is considerable, and would obviously have to be taken into account in any quantitative treatment of the behaviour of acetic acid in solutions of appreciable ionic concentration.

Table 3

Dissociation constant of acetic acid in salt solutions

Salt molality	$10^5 K_c$ in KCl	$10^5 K_c$ in SrCl$_2$
0	1·74	1·74
0·01	1·86	1·92
0·05	2·19	2·47
0·1	2·69	3·09
0·2	2·95	3·47
0·5	3·17	4·07

The increase in K_c shown in Table 3 can be expressed in terms of the activity coefficients of the ions concerned. If we insert these activity coefficients as correction factors in the equilibrium constant expression for acetic acid (HOAc) we obtain:

$$\frac{[H^+][OAc^-]}{[HOAc]} \cdot \frac{f_{H^+} f_{OAc^-}}{f_{HOAc}} = K_a = K_c \frac{f_{H^+} f_{OAc^-}}{f_{HOAc}}, \quad (20)$$

where K_c is the ordinary dissociation constant in terms of concentrations, and K_a is the thermodynamic constant. K_a is dependent only upon the temperature, and any displacement of the equilibrium by changes of salt concentration is taken care of by the variations of the activity coefficients. Experience shows that changes of ionic strength affect the thermodynamic properties of non-electrolytes in solution much less than those of electrolytes, and it is therefore a fairly good approximation to put $f_{HOAc}=1$ in equation (20). If we then take logarithms and use equation (19) the result is:

$$\log_{10} K_c = \log_{10} K_a + I^{\frac{1}{2}}/(1+I^{\frac{1}{2}}). \tag{21}$$

This equation predicts an increase of dissociation constant with ionic strength of the order of magnitude given in Table 3, and will be approximately valid for the dissociation constant of any uncharged acid in water in the neighbourhood of room temperature. An increase of this kind is termed a *positive salt effect*. If the acid is a moderately strong one, the ions produced by its own dissociation may produce significant changes in ionic strength even when no electrolyte is added to the solution. For example, dichloracetic acid has $K_a = 5.5 \times 10^{-2}$, but the values of K_c in solutions 0·01 N, 0·05 N and 0·10 N in acid are respectively 6.7×10^{-2}, 7.9×10^{-2}, and 8.6×10^{-2}.

The variation of K_c with ionic strength will of course also affect the hydrogen-ion concentration in *buffer solutions*. This obviously applies to the addition of neutral salts to a buffer solution, but it must also be remembered that the ionic constituents of the buffer also contribute to the ionic strength, which involves a summation over all ions present in the solution. This means that the hydrogen-ion concentration in a buffer solution is not dependent only on the buffer ratio, but also to some extent on the buffer concentration (cf. equation (11)). For example, equation (21) predicts that if a buffer containing 0·1 N acetic acid and 0·1 N sodium acetate is diluted with an equal volume of water, its pH changes from 4·52 to 4·58, i.e. a change of 11% in the hydrogen-ion concentration. A neglect of changes of this kind led to many spurious interpretations in the early days of the ionic theory.

So far we have considered only the dissociation of uncharged acids, and the position is modified quantitatively when acids of other charge-types are considered. A particularly simple case is presented by cation acids, such as NH_4^+. If the hydrogen-ion concentration is governed by the equilibrium $NH_4^+ \rightleftharpoons NH_3 + H^+$, as in a solution of an ammonium salt (with or without the addition of ammonia to form a buffer solution), then the mass law equation is:

$$\frac{[H^+][NH_3]}{[NH_4^+]} \cdot \frac{f_{H^+} f_{NH_3}}{f_{NH_4^+}} = K_a = K_c \frac{f_{H^+} f_{NH_3}}{f_{NH_4^+}}. \tag{22}$$

We can again take $f_{NH_3} \sim 1$, and if equation (19) is valid, we have $f_{H^+}/f_{NH_4^+} = 1$, since the activity coefficient depends only on the charge on the ion. This means that $K_c = K_a$, independent of ionic strength, and the hydrogen-ion concentration in solutions of this kind is therefore little affected by changes in ionic strength (zero salt effect). This treatment can easily be generalized for an acid of any charge type. If the acid has z positive charges, the corresponding base must have $z - 1$, and the application of equation (19) gives:

$$\log_{10} K_c = \log_{10} K_a - (z-1)I^{\frac{1}{2}}/(1 + I^{\frac{1}{2}}). \tag{23}$$

Equation (23) can be used to predict the effect of ionic strength on dissociation constants and hydrogen-ion concentrations in other systems; for example, in the equilibrium $[Fe(H_2O)_6]^{+++} \rightleftharpoons [Fe(OH)(H_2O)_5]^{++} + H^+$, $z = +3$, and the equation shows that the hydrogen-ion concentration will be *decreased* by an increase of ionic strength (negative salt effect).

A similar treatment can be applied to the effect of salt concentration on the hydroxide-ion concentration in acid-base systems. One method is to use the equations already given for $[H^+]$, combined with the relation $K_w = [H^+][OH^-]$, remembering that K_w itself varies with ionic strength according to equation (21). Alternatively, the equilibrium involving hydroxide ions can be treated directly in terms of activity coefficients. For example, if we have a buffer solution containing known concentrations of NH_3 and NH_4^+, the equilibrium involved is $NH_3 + H_2O \rightleftharpoons NH_4^+ + OH^-$, and the value of $[OH^-]$ is given by:

$$[OH^-] = \frac{[NH_3]}{[NH_4^+]} \cdot \frac{K_w'}{K_a} \cdot \frac{f_{NH_3}}{f_{NH_4^+} f_{OH^-}}, \tag{24}$$

where K_w' and K_a are thermodynamic values for the ionic product of water and the acid constant for the ammonium ion. Putting $f_{NH_3} = 1$ and using equation (21) for the ionic activity coefficients we find:

$$\log_{10}[OH^-] = \log_{10}\frac{K_w'}{K_a} + \log_{10}\frac{[NH_3]}{[NH_4^+]} + I^{\frac{1}{2}}/(1+I^{\frac{1}{2}}), \quad (25)$$

showing a positive salt effect.

For many purposes it is necessary either to eliminate or to allow for the effect of the interionic forces, and there are several ways in which this can be done.

(a) If it is possible to keep the total ionic concentrations very low (preferably below 0·001) then the deviations from simple behaviour will be small. This procedure is obviously only rarely applicable.

(b) If the ionic strength is not too high, approximate activity coefficients can be calculated from equation (21), or some similar standard expression. Provided that no multiply charged ions are present this method will give fairly reliable results up to about $I = 0·1$, though individual deviations may occur.

(c) A more generally useful procedure is to carry out a series of investigations at *constant ionic strength*, if necessary by adding appropriate amounts of a neutral salt. According to equation (21) all the activity coefficients will remain constant throughout the series, and the same is therefore true of values of K_c and K_w. The simple equations of Chapter 2 will apply, with the values of K_c and K_w appropriate to the ionic strength chosen. For many purposes it is not necessary to know these values exactly provided that their constancy can be assumed. As an example of the usefulness of experiments at constant ionic strength we give in Table 4 some data on the hydrogen-ion concentration in a series of formate buffers, determined by a catalytic method (cf. Chapter 6). In the absence of interionic effects all the solutions in Table 4 would have the same hydrogen-ion concentrations. Actually the first five entries in the Table show the expected positive salt effect, the hydrogen-ion concentration increasing by about 40% between $I = 0·01$ and $I = 0·1$. By contrast, in the last four solutions $[OH_3^+]$ remains effectively constant in spite of large variations in the buffer concentration; this is because the solutions are made up to a constant ionic strength of 0·1 by adding the appropriate amount of sodium chloride.

It is not always convenient to maintain the total ionic strength constant, and an alternative procedure (though a less desirable one) is to 'swamp out' variations by adding to the system a large constant concentration of a neutral salt. Examination of equation (18) shows that $d \log_{10} f_z/dI$ becomes numerically smaller as I increases, and variations in ionic strength should therefore be less important when

Table 4

Hydrogen-ion concentrations in formate buffers at 20°C.

$[HCOOH]/[HCOO^-] = 2·96$ throughout

[HCOOH]	[HCOO⁻]	[NaCl]	I	$10^3[OH_3^+]$
0·0296	0·0100	—	0·011	6·62
0·0592	0·0200	—	0·020	6·93
0·1480	0·0500	—	0·050	8·00
0·2220	0·0750	—	0·075	8·70
0·2960	0·1000	—	0·100	9·43 ⎫
0·1776	0·0600	0·0400	0·100	9·35 ⎬ Mean
0·0987	0·0333	0·0667	0·100	9·40 ⎭ 9·46
0·0222	0·0075	0·0925	0·100	9·65

the total ionic strength is high. However, equation (18) is not valid at large values of I, and the method is not always reliable.

In general there is no satisfactory theory of *more concentrated electrolyte solutions*, and the observed thermodynamic behaviour does not depend only on the ionic charge and the ionic strength. General equations such as (18) and (19) are not applicable, and it is no longer allowable to take the activity coefficient of uncharged molecules as unity, independent of salt concentration. The behaviour of any particular system depends in a specific manner both on the nature of the ions being considered, and on the nature of the other ions present in the solution. This is illustrated in Fig. 5, which shows how the ionic product of water, $[OH_3^+][OH^-]$, varies with ionic strength in a number of concentrated salt solutions. The data are taken from electrometric measurements by H. S. Harned and his collaborators (1933 onwards).

In describing acid-base properties in solutions of high ionic strength, a useful simplification is provided by the *acidity function* of L. P. Hammett (1934–40). Suppose that a solution (which may

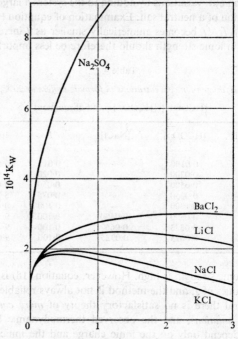

Fig. 5. Ionic product of water in salt solutions at 25°C.

have a high concentration of hydrogen ions and other ions) contains a small quantity of an uncharged base B. The equilibrium between B and BH^+ is then given by the expression:

$$\frac{[B][H^+]}{[BH^+]} \cdot \frac{f_B f_{H^+}}{f_{BH^+}} = K_a, \tag{26}$$

where K_a is the thermodynamic acid constant of BH^+.

The activity coefficients cannot be predicted if the solution is concentrated, but since B and BH^+ differ only by a proton it may be expected that the ratio f_B/f_{BH^+} will be independent of the nature of B. If this is so the quantity h_0, defined by:

$$h_0 = K_a \frac{[BH^+]}{[B]} = [H^+]\frac{f_{H^+}f_B}{f_{BH^+}}, \tag{27}$$

will be a property of the solution, independent of the nature of B and of its concentration (supposed small). The ratio $[BH^+]/[B]$ can be measured easily if the substance is an indicator, and h_0 can be calculated if a value for K_a can be obtained from measurements in dilute solutions of acids, or by other means. For a series of bases of the same charge and similar structures it is found that h_0 is to a good approximation independent of B, and may thus be used as a measure of the acidity of the solution. Conversely, an acid solution of known h_0 can be used to measure K_a for a new acid-base system BH^+—B, and this method has been widely used for the investigation of weak bases.

For some purposes it is more convenient to use a quantity H_0 defined by:

$$H_0 = -\log_{10} h_0 = pK_a + \log_{10} [B]/[BH^+]. \tag{28}$$

H_0 is termed the *acidity function*, and becomes equal to pH in dilute solution. Figure 6 (taken from indicator measurements with primary aromatic amines by L. P. Hammett) shows how H_0 varies with concentration in concentrated solutions of a number of strong acids. It is clear that h_0 increases much more rapidly than $[H^+]$, and that it depends specifically on the nature of the acid. Although no exact theory can be given for the concentration dependence of h_0, it is probable that its rapid increase in concentrated solutions (corresponding to about a thousand-fold increase of acidity between 1M and 10M acid) is closely connected with the hydration of the hydrogen ion. If the latter is written as $H(H_2O)_n^+$, then (neglecting the hydration of other species) the protonation of a base can be represented by the equation $B + H(H_2O)_n^+ \rightleftharpoons BH^+ + nH_2O$, and the degree of protonation will be greatly decreased by the reduction of water activity in concentrated solutions. It has in fact been shown (P. A. H. Wyatt, R. P. Bell, 1957) that the observed values of H_0 are closely related to the water activity, as measured by its partial vapour pressure, and that over a considerable concentration range the results can be interpreted in terms of the species $H_9O_4^+$, already mentioned in Chapter 1.

It was originally hoped that the acidity function, although it might depend on the charge on the base (as represented by the symbols H_0, H_-, H_+ etc.), would be largely independent of its chemical

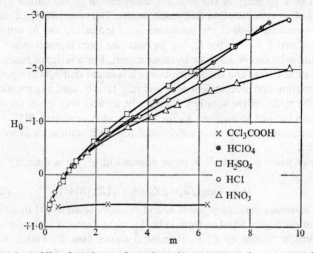

Fig. 6. Acidity functions of moderately concentrated aqueous acids. (From L. P. Hammett: 'Physical Organic Chemistry', *McGraw-Hill.*)

structure. Detailed investigation has shown that this simple hypothesis represents only a very crude approximation, and in consequence the literature now contains a bewildering array of acidity functions (for example, H_0', H_0'', and H_0''' for primary, secondary, and tertiary amines), each strictly applicable only to a restricted range of compounds. As a result, the whole concept of the acidity function has lost some of its orginal attraction. It is, however, still used widely in connection with the dependence of reaction rates on acidity, to which reference will be made in Chapter 6.

Little can be said about the effect of interionic attraction in *non-aqueous solutions*, except that it is usually much greater than in water because of the lower dielectric constant. The theoretically derived constant A in equation (18) contains the factor $1/\varepsilon^{3/2}$: this gives some measure of the effect of decrease in dielectric constant, and has been approximately verified for solutions in methanol and ethanol. However, the range of validity of such theoretical expressions is much less than in water, and different electrolytes show markedly individual behaviour. It is for this reason that the interpretation of acid-base properties in non-aqueous solutions often rests on at best a semi-quantitative basis.

GENERAL REFERENCES

D. A. MacInnes, *The Principles of Electrochemistry* (New York, 1939).

R. A. Robinson and R. H. Stokes, *Electrolyte Solutions* (London, 1959).

R. P. Bell, *The Proton in Chemistry* (Ithaca, N.Y., and London, 1959), Chapter VI.

'Interactions in Ionic Solutions', *Disc. Faraday Soc.*, 1957, **24**.

M. A. Paul and F. A. Long, 'H_0 and Related Indicator Acidity Functions', *Chem. Reviews*, 1957, **57**, 1.

E. J. King, *Acid-Base Equilibria* (Oxford, 1965), Chapter 12.

G. Kortüm, *Treatise on Electrochemistry* (2nd edn.), (Elsevier, 1965) Chapter 5.

Acid-base Strength and Molecular Structure

The correlation of acid-base strengths with molecular structure has been used in a wide variety of ways to test theories as to the nature of chemical bonds and the distribution of electrons in molecules. Measurements of acid-base strength do in fact constitute the only large body of comparable data which are available for studying the effect of structure on equilibrium constants, and from the theoretical point of view equilibrium data are easier to interpret than are the results of reaction velocity measurements. (Similar information can sometimes be derived from thermochemical measurements, but the effects sought frequently represent a very small fraction of the observed quantities.) The subject is a very extensive one, and in this chapter we shall give only a limited number of examples of the way in which measured acid-base strengths have been interpreted.

There is one difficulty which occurs in all attempts to correlate measured thermodynamic quantities with the predictions of a molecular model. The theoretical calculations (whether based on quantum theory or on a simplified electrostatic picture) give information about the energy of isolated molecules at absolute zero. Experimental data, on the other hand, relate to a finite temperature, and (in the present instance) to molecules in solution. The measured thermodynamic quantities represent averages over a large number of molecules, which differ individually in their thermal energies (translational, rotational, and vibrational) and in their interactions with the solvent, and it is by no means certain that these average values will be affected by structural factors in the same way as the hypothetical values derived from the model.

The two most important thermodynamic quantities concerned are ΔH, the change in heat content, and ΔG°, the standard free energy change. They both have the dimensions of energy, and represent different ways of averaging the energies of the individual molecules concerned. Measured dissociation constants are related directly to the free energy change by the thermodynamic equation:

$$-\Delta G^{\circ} = RT \log K, \tag{29}$$

while the change in heat content can be determined either calorimetrically or from the free energy by the equation:

$$\Delta H = \Delta G^{\circ} + T \Delta S^{\circ} = \Delta G^{\circ} - T \left(\frac{\mathrm{d}\Delta G^{\circ}}{\mathrm{d}T} \right)_{p}, \tag{30}$$

where ΔS° is the standard entropy change of the reaction. At absolute zero the distinction between free energy and total energy disappears, but at a finite temperature it is impossible to say *a priori* which of the two quantities ΔG° and ΔH will be more closely related to the properties of the model. An optimistic view would be to assume that changes in ΔG° are always accompanied by parallel changes in ΔH (corresponding to no variation in the entropy change), but experiment shows that this is by no means the case. Table 5 gives a selection of data (collected by D. H. Everett and W. F. K. Wynne-Jones, 1939) for carboxylic acids in water at 25°. All values are given relative to formic acid, since the values of ΔG° and ΔS° are then independent of the units in which the concentrations are expressed.†

It will be seen that the values of $T\Delta S^{\circ}$ vary widely and erratically, so that there is no simple correlation between ΔG° and ΔH. There has been much discussion as to which of these quantities is best suited for comparision with molecular models. Some theoretical arguments have been advanced in favour of ΔG° (M. G. Evans and M. Polanyi, 1936), and the problem is often solved in practice by the lack of data for ΔH. In this chapter we shall follow the usual practice of drawing conclusions from values of K at a single temperature (i.e. from ΔG°). This is certainly safe in dealing with changes in K of many powers of ten, but there is need for caution in interpreting small changes of K, even when very similar acids are involved. Sometimes the entropy changes themselves can be related to structural factors, for example the extent of solvation of the acid or basic species.

---

† The ordinary dissociation constant $[R.COO^-][OH_3^+]/[R.COOH]$ has the dimensions of concentration, while the value of $[R.COO^-][H.COOH]/[R.COOH][H.COO^-]$ is independent of concentration units in dilute solutions.

Table 5

Thermodynamic data for the reaction
$R.COOH + H.COO^- \rightleftharpoons R.COO^- + H.COOH$ *in water at 25°C.*

Acid	pK of acid	k.cal/mole		
		$\Delta G°$	ΔH	$T\Delta S°$
Propionic	4·87	1·53	−0·13	−1·66
Butyric	4·82	1·46	−0·66	−2·12
Acetic	4·76	1·37	−0·06	−1·43
Anisic	4·48	1·00	+0·84	−0·16
Cinnamic	4·41	0·90	+0·64	−0·26
p-Toluic	4·34	0·81	+0·34	−0·47
m-Toluic	4·24	0·68	+0·11	−0·57
Benzoic	4·17	0·57	+0·08	−0·49
m-Hydroxybenzoic	4·08	0·45	+0·21	−0·24
o-Toluic	3·88	0·17	−1·36	−1·53
m-Iodobenzoic	3·86	0·15	+0·23	+0·08
Lactic	3·86	0·15	−0·13	−0·28
Glycollic	3·83	0·11	+0·14	+0·03
Formic	3·79	0	0	0
m-Nitrobenzoic	3·45	−0·41	+0·36	+0·77
Salicylic	2·97	−1·05	+1·04	+2·09
o-Chlorobenzoic	2·88	−1·18	−2·43	−1·25
o-Iodobenzoic	2·86	−1·21	−3·21	−2·00
Monochloroacetic	2·86	−1·21	−1·11	+1·10
o-Nitrobenzoic	2·19	−2·13	−3·32	−1·19
α-Bromocinnamic	1·98	−2·41	−2·95	−0·51

We shall first consider the acidities of some *simple hydrides*, and some approximate pK-values are given in Table 6, which is arranged according to the periodic table. The values relate to aqueous solutions, but many of the high or low ones are obtained by measurements in other solvents or by theoretical estimation.

TABLE 6

Acid strengths (pK-values) of simple hydrides

CH_4 (58)	NH_3 (35)	OH_2(16)	FH (3)
	PH_3 (27)	SH_2(7)	ClH (−7)
		SeH_2(4)	BrH (−9)
		$TeH(3)_2$	IH (−10)

Table 6 shows obvious trends, but their interpretation in terms of atomic and molecular structure is less simple. The dissociation of an acid HX in solution can be split up into the following steps:

1. HX (solution) $\rightarrow HX$ (gas).
2. $HX \rightarrow H + X$ (gas).
3. $H \rightarrow H^+ + \varepsilon$ (gas).
4. $X + \varepsilon \rightarrow X^-$ (gas).
5. H^+ (gas) $\rightarrow H^+$ (solvated).
6. X^- (gas) $\rightarrow X^-$ (solution).

Of these steps, (3) and (5) are the same for all acids, while the energies involved in (1) are small. The variations in the acidity of simple hydrides thus depend chiefly on variations in the following terms:

(i) The energy of dissociation of the hydride to give a hydrogen atom.
(ii) The electron affinity of the atom or radical remaining.
(iii) The energy of solvation of the anion formed.

In the series CH_4—HF the increasing electron affinity of the atom or radical is the chief factor determining the increase in acidity, but in general all three factors are involved. For example, in the series of hydrogen halides the dissociation energies will lead to an increase of acidity with increasing atomic weight, but the solvation energies and electron affinities will tend in the reverse direction. At present we have insufficient knowledge of the quantities concerned to permit of any detailed discussion, and we shall only mention here that the 'weakness' of hydrogen fluoride is not necessarily anomalous, as is sometimes supposed.† Its pK fits well into the series CH_4—NH_3—OH_2—HF, and in the series of halogen hydrides the distinction between 'weak' and 'strong' acids is an arbitrary one. It is true that the difference between HF and HCl appears to be unexpectedly large, but it must be remembered that the absolute pK-values for HCl, HBr, and HI are very uncertain: some workers have preferred a

† It should be stressed that the value 7×10^{-4} for the dissociation constant of HF in water refers to the simple equilibrium $HF + H_2O \rightleftharpoons OH_3^+ + F^-$, due allowance having been made for the formation of HF_2^-. On the other hand, the acidic properties of anhydrous hydrogen fluoride (cf. Chapter 3) are much more powerful than would be anticipated from its behaviour in water, and this anomaly is probably related to the highly associated nature of the pure liquid.

value of -3 for HCl, and all that can be stated with certainty is that pK is negative, increasing numerically in the series HCl, HBr, HI.

It should be noticed that there is no general relation between acidity and the electronegativity of the atoms concerned. The series CH_4—HF corresponds to an increasing dipole moment of the link X—H, but in the series HF—HI an increase in acidity is accompanied by a decrease of dipole moment, and the same is true of the series H_2O—H_2Te. This is because, as already stated, the electron affinity of the atom X is only one of the factors determining acidity, there being no general relation between the ionic character and total strength of a bond.

A large proportion of the acids about which we have quantitative information belong to the class of *oxy-acids* containing the group —OH. This class of course includes the organic carboxylic acids and phenols, but for the present we shall consider only the simple inorganic acids with the general formula H_mXO_n, where m and n depend upon the valency of X. These acids cover a very wide range of strengths, and it is a striking fact that the most important factor in determining acidity is the *number of equivalent oxygen atoms in the anion*, it being assumed for this purpose that all oxygen atoms not carrying a hydrogen atom are equivalent. A selection of values is given in Table 7, and it will be seen that they fall into fairly well-defined groups, though there are naturally individual differences according to the nature of the central atom.

It is interesting to note that phosphorous and hypophosphorous acids fit into the above scheme only if they are given the formulae $HPO(OH)_2$ and $H_2PO(OH)$ rather than $P(OH)_3$ and $HP(OH)_2$ respectively: this is of course in accordance with other evidence as to their structure. In other cases the degree of hydration of the acid is important. The value for H_2CO_3 takes into account the small proportion of hydrated carbon dioxide in solution, while telluric acid must be $Te(OH)_6$ rather than $TeO_2(OH)_2$. It is surprising that boric acid fits so well into the series, since there is good evidence that the borate ion in aqueous solution has the formula $B(OH)_4^-$, so that the dissociation process is $B(OH)_3 + H_2O \rightleftharpoons B(OH)_4^- + H^+$, rather than $B(OH)_3 \rightleftharpoons B(OH)_2O^- + H^+$.

Considerations of the above kind can be extended to the successive dissociation constants of polybasic acids, and also to more complicated acid molecules such as $H_4P_2O_7$. It is in fact possible to formulate quantitative rules for predicting approximately the dissociation constants of a wide variety of acids (e.g. J. E. Ricci, 1948), but we

Table 7

Acid strengths of oxy-acids

Acid	pK		Number of equivalent oxygen atoms in anion
ClOH	7·3		1
BrOH	9		1
IOH	10	Mean	1
B(OH)₃	9·2	9·3	1
Al(OH)₃	11·2		1
As(OH)₃	9·2		1
Te(OH)₆	8·7		1
ClO(OH)	2		2
NO(OH)	3·4		2
SO(OH)₂	1·8		2
CO(OH)₂	3·4	Mean	2
PO(OH)₃	2·1	2·3	2
AsO(OH)₃	2·3		2
H₂PO(OH)	1·1		2
HPO(OH)₂	2·0		2
NO₂(OH)	−1		3
ClO₂(OH)	−1	Mean	3
SO₂(OH)₂	−3	−1·7	3
ClO₃(OH)	−8(?)		4

shall not discuss these here as their basis is doubtful. The increase in acidity with the number of equivalent oxygen atoms is usually attributed to a *resonance stabilization* of the anion (L. Pauling, 1939): for example, the acid sulphate ion can be written as a resonance hybrid of the three following forms:

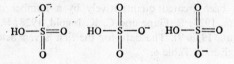

(not taking into account other formulations involving semi-polar double bonds), while in an ion such as ClO^- or $H_2BO_3^-$ all the charge is concentrated upon a single oxygen atom. It may be felt that 'resonance' is a somewhat artificial way of describing the inadequacy of conventional structural formulae to represent the equivalence of the oxygen atoms in these ions. If this equivalence is accepted as an experimental fact (established by X-ray or similar methods), then the

Table 8

Dissociation constants of dicarboxylic acids at 25°C.

Acid		pK_1	pK_2	r (calc.) (Å)
Oxalic	COOH.COOH	1·24	5·17	0·8
Malonic	COOH.CH$_2$.COOH	2·75	5·36	1·5
Succinic	COOH.(CH$_2$)$_2$.COOH	4·13	5·35	5·0
Glutaric	COOH.(CH$_2$)$_3$.COOH	4·34	5·27	9·2
Adipic	COOH.(CH$_2$)$_4$.COOH	4·41	5·28	11·5
Pimelic	COOH.(CH$_2$)$_5$.COOH	4·48	5·31	13·2
Suberic	COOH.(CH$_2$)$_6$.COOH	4·51	5·33	14·5
Azelaic	COOH.(CH$_2$)$_7$.COOH	4·55	5·33	16·8

increase in acid strength follows as a consequence of the delocalization of electrons over several bonds and atoms, which will stabilize the anion.

The next important factor in determining acid strength is that of *electrical charge*. Other things being equal, we shall expect an ion with a negative charge to lose a proton less readily than a neutral molecule, and a positively charged ion more readily. This effect certainly plays a part in determining the successive dissociation constants of oxy-acids, e.g. H_3PO_4 (pK 2·1), $H_2PO_4^-$ (pK 7·2), $HPO_4^=$ (pK 11·9), but here it is complicated by other factors, such as the spread of charge over several oxygen atoms. The clearest case of a charge effect is in the two dissociation constants of *dicarboxylic acids*, which has been treated quantitatively by a number of authors (N. Bjerrum, 1923; R. Gane and C. K. Ingold, 1928; F. H. Westheimer *et al.*, 1938–9). The data for the first seven acids of this series are given in Table 8.

It will be seen that the dissociation constant of the singly charged anion is at first smaller than that of the original acid, and that the difference between the successive pK- values decreases fairly steadily as the length of the chain increases. A part of this difference depends on a purely statistical effect. In the first place the acid molecule contains two dissociable protons compared with only one in the singly charged anion, and in the second place the singly charged anion has only one point at which a proton can be accepted, compared with two such points in the doubly charged anion. This means that in comparing the equilibrium constants of the two processes:

(1) $COOH-----COOH \rightleftharpoons COOH-----COO^- + H^+$

and (2) $COOH-----COO^- \rightleftharpoons COO^------COO^- + H^+$,

the first should be four times as great as the second on purely statistical grounds.†

The observed ratios are all greater than four, and it is this excess which can be attributed to the effect of the charge. The simplest method of estimating this effect is to consider the repulsion of the two negative charges in the doubly charged anion: this corresponds to a potential energy of $e^2/\varepsilon r$, where e is the electronic charge, ε the dielectric constant, and r the distance between the two charges. No similar electrostatic term occurs for any of the other species concerned, and by the Boltzmann principle the effect of charge is thus to multiply the second dissociation constant by a term $\exp(-e^2/\varepsilon rkT)$. This is superimposed upon the statistical effect, so that the simple theory predicts:

$$\ln \frac{K_1}{4K_2} = \frac{e^2}{\varepsilon rkT}. \tag{31}$$

The last column of Table 8 shows the values of r calculated by inserting the measured values of K_1 and K_2 in equation (31), taking $\varepsilon = 80$. They are of the expected order of magnitude, but are too small

† The argument outlined here is essentially kinetic in nature, depending on considerations about the probability of the dissociation and association processes. A more satisfactory (but more sophisticated) method appeals only to equilibrium considerations: in calculating these equilibrium constants by the methods of statistical mechanics the factor four appears as the product of two *symmetry numbers*, two for the acid molecule and two for the doubly charged anion.

for the lower members of the series. The weakest point of the theory is the use of the macroscopic dielectric constant of the solvent for distances of molecular dimensions. Better agreement is obtained by using a 'local' dielectric constant smaller than the macroscopic value, but the calculation of such dielectric constants is itself a speculative undertaking, and we shall not describe these refinements here. A similar treatment can of course be applied to many related problems, for example geometrical isomers (maleic and fumaric acids, etc.), diamines, and the high acidity of species like $[Fe(H_2O)_6]^{+++}$ ($pK = 2.7$).

We shall now consider the effects of *substituents in organic acids*, especially carboxylic acids. This is a vast subject, and only a limited number of examples will be given here. In treating this kind of problem attention is often focussed on the actual process of dissociation, with a view to determining whether this process will be promoted or inhibited by the substituent considered. This kind of treatment is necessary when the *velocities* of protolytic reactions are being considered, but it is not the simplest way of treating *equilibria*, which will involve the effect of the substituent on both the forward and the reverse action. A safer method is to consider the effect of the substituent on the energy of the acid and of its corresponding base (normally the anion), and we shall follow this procedure.

The aliphatic carboxylic acids have $pK \sim 5$, and are thus much more strongly acidic than the alcohols (pK 16–19). This is undoubtedly because the negative charge in the anion is divided equally between the two oxygen atoms: we have seen in dealing with the inorganic oxy-acids that such a spread of charge (whether or not interpreted as resonance between two structures) brings about a stabilization of an anion. In many instances this charge distribution is not changed by substituents, and is therefore not relevant to their effect.

The simplest type of substituent effect involves a substituent bearing a *net electrostatic charge*. One example of this is the increase in pK produced by the group —COO^-, already considered, and a corresponding decrease in pK is produced by introducing a positively charged group such as —NMe_3^+. For example, $NMe_3^+CH_2COOH$ has pK 1.83, compared with 4.75 for acetic acid: this is clearly due to the stabilization of the species $\overset{+}{N}Me_3.CH_2.COO^-$ by the attraction between the positive and negative charges. However, it is not always safe to assume that the effect of a substituent is determined, even qualitatively, by its net charge. An interesting example is provided by the following series of pK values (R. P. Bell and G. A. Wright, 1961),

CH$_3$COOH 4·75,   SO$_3^-$(CH$_2$)$_3$COOH 4·91,

SO$_3^-$CH$_2$COOH 4·20,   SO$_3^-$(CH$_2$)$_4$COOH 3·04,

SO$_3^-$(CH$_2$)$_2$COOH 4·74,

which indicate that the negatively charged sulphonate group makes the anion less stable when it is separated from the carboxylate group by several carbon atoms, but stabilizes it when only one CH$_2$-group intervenes. This behaviour can be rationalized by supposing that (at least in the anion) the structure of the sulphonate group can best be written as

i.e., without any multiple bonds between sulphur and oxygen. At points close to the sulphur atom the effect of its double positive charge will outweigh that of the three negative charges on the more distant oxygen atoms, but at greater distances the effect of the latter will become more important, and the group will behave as a single negative charge.

Another simple type of substituent effect is an *electrostatic dipole effect* exemplified by halogen substitution. Table 9 gives the pK-values for some chlorinated aliphatic acids, the effect of bromine and iodine substitution being similar.

The substitution of hydrogen by chlorine always causes an increase of strength, which is accentuated by the presence of more than one chlorine on the same carbon atom, but which dies away rapidly as the point of substitution recedes from the carboxyl group. This effect can be explained simply by the stabilization of the anion through interaction between the C—Cl dipole and the negative charge. For

Table 9

Acid strengths of some chloro-acids

Acid	pK	Acid	pK
CH₃.COOH	4.76	CH₂Cl.CH₂COOH	4·07
CH₂Cl.COOH	2.81	CH₃.CH₂.CH₂.COOH	4·82
CHCl₂.COOH	1.3	CH₃.CH₂.CHCl.COOH	2·86
CCl₃.COOH	0.7	CH₃.CHCl.CH₂.COOH	4·05
CH₃.CH₂.COOH	4.85	CH₂Cl.CH₂.CH₂.COOH	4·52
CH₃.CHCl.COOH	2.83		

example, the charge distribution in the anion of monochloroacetic acid can be represented as follows:

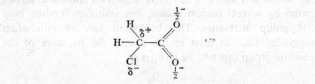

where $\cdots$ represents the bond of order intermediate between single and double, and δ^+ and δ^- are the fractional charge displacements of the C—Cl dipole, the much smaller C—H dipoles being omitted. The energy of this structure will be lowered by the attraction between δ^+ on the carbon atom and the negative charge on the oxygen atoms, and since there is no interaction of comparable magnitude in the undissociated acid molecule, the result is an increased dissociation. It is sometimes objected that this explanation does not account for the great strength of trichloroacetic acid, since the resultant dipole moment of the group —CCl₃ must be almost the same as that of $\diagdown$CCl. This certainly implies that the two groups will exert similar fields at a large distance, but the same is not true for distances comparable to the dimensions of the dipole: here the magnitude of the charge on the carbon atom is the important factor, and this will certainly be greater in —CCl₃ than in $\diagdown$CCl.

Several other groups will increase the strength of carboxylic acids by a similar effect, e.g. —OH (CH₂OH.COOH, pK 3·8), —CN

$(CH_2CN.COOH, pK 2\cdot45)$ and $-NO_2$ $(CH_2NO_2.COOH, pK 1\cdot75)$. In the first the link O^--C^+ has the polarity shown, while in the last two the positive end of the group dipole is nearest to the carboxyl group. No cases are known in which a simple substitution causes a large decrease in the strength of a carboxylic acid.

There are very many substituents whose effect cannot be explained in terms of dipoles, but in some instances other types of electrostatic effect can be invoked. For example, the substitution of *phenyl* into an aliphatic acid produces an increase in strength (CH_3COOH, $pK 4\cdot75$; $C_6H_5CH_2COOH$, $pK 4\cdot31$; $(C_6H_5)_2CH.COOH$, $pK 3\cdot94$; $CH_2.CH_2.COOH$, $pK 4\cdot85$; $C_6H_5.COOH$, $pK 4\cdot37$). The phenyl group has no strong polarity, but it is easily polarizable in the field of the anion, whose energy it will thus lower. It is a matter of taste whether we regard this high polarizability as an experimental fact (e.g. from the refractive indices of aromatic compounds) or as a consequence of possible structures for the phenyl group such as:

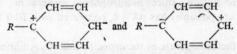

the former of which would be important if R bears a negative charge.

There are many instances, however, in which the influence of phenyl and similar substituents cannot be explained in terms of any general electrostatic effect, but demands a consideration of particular kinds of electronic shifts in the bonds concerned (*mesomerism* or *resonance*). For example, *phenol* ($pK 10$) is much more strongly acidic than the alcohols ($pK 16-19$), and this difference is too large to be explained by the polarizability of the phenyl group. It is probably due to the partial distribution of charge in the phenate ion, which can be represented by saying that the structure of the ion is not given simply by (I) below, but that other structures such as (II) and (III) participate to some extent.

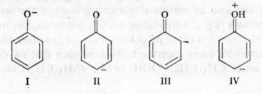

I   II   III   IV

In the actual state of the ion a part of the negative charge will be transferred from the oxygen atom to the *ortho*- and *para*-positions in the ring, and this spread of charge corresponds to a lower energy and a greater ionization. Similar structures (e.g. IV) are formally possible in the undissociated molecule, but in this case they involve the separation of positive and negative charges and will not materially lower its energy. Substituents in the *ortho*- and *para*-positions may also take part in such mesomerism: for example, *o*- and *p*-nitrophenol both have pK 7·2 because of the possibility of electron distributions represented by:

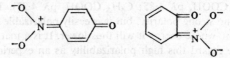

The effect of the nitro-group must however be partly of a simple electrostatic nature, since *m*-nitrophenol (where no mesomeric structures are possible) has pK 8·4, compared with 10·0 for phenol itself.

Similar factors operate in the *aromatic amines*. Aniline is a much weaker base than methylamine, or, in other words, the ion $C_6H_5.NH_3^+$ (pK 4·7) is a much stronger acid than $CH_3NH_3^+$ (10·7). Again it is the conjugate base which is stabilized by the structures:

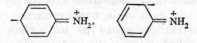

which are not possible when a proton has been added on to the nitrogen atom. This kind of charge distribution is verified by comparing the dipole moments of the aromatic and aliphatic amines, the former being the larger (L. E. Sutton, 1931). The effect is accentuated by the presence of two phenyl groups (e.g. $(C_6H_5)_2NH_2^+$ has pK 0·8) and also by certain mesomeric groups in the *ortho*- and *para*-positions (e.g. *o*-nitroaniline, pK 0·1, *p*-nitroaniline, pK 1·9).

The interpretation of the pK of *benzoic acid* (4·2) is more subtle (and correspondingly more uncertain). We have seen that pK decreases in the series $CH_3.CH_2.COOH$ (4·85), $C_6H_5.CH_2.CH_2.COOH$

(4·66), $C_6H_5.CH_2.COOH$ (4·31), and it would be anticipated that benzoic aicd would be considerably more acidic. The factor opposing an increase in strength is probably the stabilization of the undissociated acid by structures such as (I) below: the corresponding structure (II) for the anion is much less likely as it involves an increase of the charge density on two oxygens of the carboxyl group.

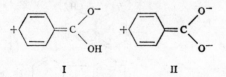

I   II

This explanation receives some support from the fact that all *ortho*-substituted benzoic acids are stronger than benzoic acid itself, independent of the nature of the substituent group. This may be because the substituent prevents the carboxyl group from assuming the planar configuration implied by (I), thus lowering the stability of the undissociated molecule. It must be admitted, however, that these explanations are ingenious rather than convincing.

It would be possible to give many other examples of the effect of mesomerism upon acidity, for example in heterocyclic compounds, in amides and in amidines. Instead we shall now consider *acids with the acidic hydrogen attached to carbon*. All the organic acids so far considered have contained the groups —OH or $>$NH, and $->$CH does not as a rule exhibit any acidic properties. It has been estimated that methane has $pK \sim 58$, which means that it will not in practice behave as an acid. Ethylene and acetylene are presumably somewhat more acidic, since the latter forms some metallic derivatives and exchanges deuterium slowly with D_2O in strongly alkaline solution; the pK of acetylene has been estimated as about 26. Much greater acidic power appears in more complex molecules, where the charge of the anion does not reside wholly on the carbon atom which has lost the proton, but is transferred to other atoms by electronic rearrangements within the molecule. Some examples are given in Table 10.

In the three hydrocarbons in Table 10 the anion is stabilized by

Table 10

Acid strengths of some carbon acids

(a) Hydrocarbons

Acid	pK	Structure of acid (acidic proton asterisked)	Number of structures Acid	Number of structures Anion
Triphenylmethane	~33	$(C_6H_5)_3$ CH*	6	9
Indene	21	(indene with CH₂*)	2	9
Fluoradene	11	(fluoradene with CH*)	8	62

(b) Ketones and nitroparaffins

Acid	pK	Structure of acid	Structure of anion
Acetone	19	CH_3COCH_3	$CH_2:C(O^-).CH_3$
Acetylacetone	8·9	$CH_3COCH_2COCH_3$	$CH_3C(O^-):CH.COCH_3$ $CH_3CO.CH:C(O^-)CH_3$
Nitromethane	10·2	CH_3NO_2	$CH_2:\overset{+}{N}\overset{O^-}{\underset{O^-}{}}$
Nitroethane	8·5	$CH_3CH_2NO_2$	$CH_3CH:\overset{+}{N}\overset{O^-}{\underset{O^-}{}}$
2-Nitropropane	7·7	$(CH_3)_2CHNO_2$	$(CH_3)_2C:\overset{+}{N}\overset{O^-}{\underset{O^-}{}}$

distributing the negative charge over a number of carbon atoms. Thus for triphenylmethane anion a typical structure is

$$(C_6H_5)_2C = \left\langle \bigcirc \right\rangle -$$

and eight other structures can be written in which the charge resides on the *ortho* or *para* carbon atoms of each ring in turn. The indene anion can be written with the charge on any one of the nine carbon atoms: this is also true for the anion of fluoradene, and in this case there are also further possibilities of distributing the single and double bonds in different ways. There is clearly a relation between the strength of these hydrocarbon acids and the extent to which the negative charge is delocalized in the anion, just as we found for the inorganic oxy-acids in Table 7.

A much greater acid strength results when the charge in the anion has moved from carbon to oxygen, as in the last five examples, where the structural change can be verified by comparing the ultra-violet absorption spectra of the anion and the original substance. If these anions add on a proton at the oxygen atom, we obtain isomers of the original acids (e.g. $CH_2\!:\!C(OH).CH_3$, $CH_2\!:\!\overset{+}{N}\!\!\left\langle\begin{array}{c}O^-\\OH\end{array}\right)$ which in some cases are known as separate individuals. A. Hantzsch (1899–1924) regarded these isomers as the 'true acids' from which the anions are derived, and termed the original compounds *pseudo-acids*: he also believed that the ionization of a pseudo-acid (which is sometimes a relatively slow process) could not take place directly, but only by the intermediate formation of the 'true' acid. We still use the term pseudo-acid in this sense, but it is not now believed that its ionization involves the prior formation of the isomeric form. This question will be considered further in the next chapter.

So far nothing has been said about the effect of *alkyl groups* upon acid-base properties. The way in which alkyl groups modify the pro-perties of an organic molecule (e.g. its reactivity or heat of formation) has been a field of much theoretical discussion, but the whole sub-ject is in a far from satisfactory condition, as the same facts are

68 ACIDS AND BASES

explained by different authors in very different ways. We shall confine ourselves here to stating a few of the facts about alkyl substistution in acids and bases. Alkyl substitution in a carboxylic acid usually lowers its strength, though the effect is not a large one (e.g. CH_3COOH, pK 4·75; $CH_3.CH_2COOH$, pK 4·85; $(CH_3)_3C.COOH$, pK 5·04). In amines, the direct attachment of alkyl groups to the nitrogen atom makes the base stronger (i.e. the corresponding acid weaker), but the data as a whole show many irregularities. It is probably significant here that the introduction of alkyl groups produces large changes in the entropy of ionization, which can be related to varying degrees of solvation (D. H. Everett and W. F. K. Wynne-Jones, 1940; A. F. Trotman-Dickenson, 1949; A. G. Evans and S. D. Hamann, 1951). These facts are at least qualitatively concordant, but it is disturbing to find that in the nitroparaffins (cf. Table 10) the introduction of alkyl groups produces a large *increase* in acidity. The question is clearly a complex one, and we shall not attempt its elucidation here.

GENERAL REFERENCES

L. P. Hammett, *Physical Organic Chemistry* (New York, 1940), Chapters 2 and 9.

G. W. Wheland, *The Theory of Resonance* (New York, 1944), Chapter 7.

R. P. Bell, *The Proton in Chemistry* (Ithaca, N.Y., and London, 1959), Chapter VII.

G. Kortüm, *Treatise on Electrochemistry* (2nd edn.) (Elsevier, 1965), Chapter 10.

E. J. King, *Acid-base Equilibria* (Oxford, 1965), Chapter 7.

CHAPTER 6

The Rates of Acid-base Reactions
(Acid-base Catalysis)

In the first edition of this book (1952) the title of the present chapter was *Acid-base Catalysis*, and little reference was made to direct studies of the rates of acid-base reactions. At that date few such studies had been made, but during the last fifteen years or so much more information has become available, and it now seems logical to consider first the general problem of the rates of reactions between acids and bases, followed by acid-base catalysis as a special case of such reactions.

It is common experience that simple acid-base reactions in aqueous solution (neutralization, hydrolysis, etc.) are extremely rapid. They have often been described as 'instantaneous' though actually this only means that they are effectively complete in the time required for mixing two solutions (say 1 second). Methods have been devised, particularly by Eigen and his collaborators, for measuring the rates of very fast reactions in solution: these involve mainly the so-called *relaxation techniques*, in which mixing problems are avoided by using rapid changes of pressure, temperature or electric field to displace an equilibrium in solution.†

These methods make it possible to observe reactions which take place in periods as short as 10^{-6} or even 10^{-8} seconds, and thus to obtain quantitative velocity constants for many 'instantaneous' reactions.

Some of the results in aqueous solution show a particularly simple pattern. Thus (with a few exceptions involving steric effects, or internal hydrogen bonding) exothermic reactions of hydrogen ions with bases, involving the attachment of a proton to an oxygen or a

† For further information about these techniques, see general references at the end of this chapter.

nitrogen atom, take place with second-order velocity constants close to 4×10^{10} mole^{-1} sec^{-1}. Very similar values (though somewhat lower) are found for the reaction of hydroxide ions with acids in which the proton is attached to an oxygen or a nitrogen atom. Examples of this behaviour are given in Table 11.

The approximate constancy of the velocity constants in Table 11, over a wide range of acid strength and chemical structure, suggests

Table 11

Diffusion-controlled reactions of hydrogen and hydroxide ions in water at 15°–25°C.

Velocity constants in l mole^{-1} sec^{-1}

(a) Base $+ H_3O^+ \xrightarrow{k_1}$ Acid $+ H_2O$

Acid	pK	$10^{-10} k_1$
Ammonium	9·2	4·3
Trimethylammonium ion	9·8	2·5
Carbonic acid	3·8	4·7
Acetic acid	4·8	4·5
Benzoic acid	4·2	3·6
p-Nitrophenol	7·1	3·6

(b) Acid $+ OH^- \xrightarrow{k_1}$ Base $+ H_2O$

Ammonium ion	9·2	3·4
Phenol	10·0	1·4
Trimethylammonium ion	9·8	2·1
Piperidinum ion	11·1	2·2

that the observed rate is not limited by chemical factors, but merely represents the rate at which the reacting species can diffuse together. In fact, simple calculations based on a model of spherical molecules diffusing in a continuous medium predict a velocity constant of approximately 10^{10} l mole^{-1} sec^{-1} in water at room temperature for a reaction which takes place every time the two reactants come together. This represents the maximum possible velocity constant for second-order reactions in this medium, and reactions in this category are known as *diffusion-controlled*. In such reactions there is no chemical activation energy, though there will be a small apparent activation energy due to the fact that the viscosity of the medium decreases with increase of temperature, leading to an increase in the encounter rate.

Since in the reactions in Table 11 the equilibrium position is far to the right, the reverse reactions (i.e. acid or base $+ H_2O$) must be relatively slow, and their velocity constants k_2 will be considerably less than 10^{10} 1 mole^{-1} sec^{-1}. However, these reactions also involve no chemical energy barrier, in the sense of an energy maximum through which the reacting system passes, and the observed activation energy (apart from the small contribution from the temperature variation of viscosity) will be equal to the endothermicity. Since the equilibrium constant is given by k_1/k_2, the ratio of velocity constants, it is readily seen that for reactions in which k_1 is effectively constant k_2 will be proportional to K_a (the dissociation constant of the acid) for a series of reactions acid $+ H_2O$, and inversely proportional to K_a for the reactions base $+ H_2O$.

When oxygen or nitrogen acids or bases other than H_3O^+ and OH^- react with one another, the observed velocity constants are appreciably lower than the maximum possible values. However, they are still very high, being commonly in the range 10^8 to 10^9 1 mole^{-1} sec^{-1} for exothermic reactions, so that the energy barriers must be small. Very much smaller velocities are observed for most reactions involving *carbon acids*, i.e. species in which the acidic proton is attached to a carbon atom. These are commonly such weak acids that it is not possible to prepare aqueous solutions containing an appreciable concentration of the corresponding base, and there are therefore not many systems in which the rate of reaction can be observed directly.

The classical example of this class of reaction is the neutralization of nitro-alkanes by hydroxide ions. For example, the reaction $CH_3CH_2NO_2 + OH^- \rightarrow [CH_3CHNO_2]^- + H_2O$ at 25°C has a velocity constant of only 5·2 1 mole^{-1} sec^{-1}, compared with $1·4 \times 10^{10}$ for the analogous reaction of phenol, a somewhat weaker acid. This behaviour was first observed by Hantzsch (1899), and the marked contrast with the behaviour of most acids led him to classify nitro-alkanes and similar compounds as pseudo-acids, a term which has been widely used.†

However, it is now clear that any sharp distinction between normal acids which react rapidly and pseudo-acids which react slowly is highly arbitrary, and will in fact depend upon the experimental methods employed. For example, the velocity constant for the reaction between acetylacetone and hydroxide ions is now known to be 4×10^4 l mole^{-1} sec^{-1}: this would be regarded as an 'instantaneous' reaction as long as classical methods of observation were used, but modern techniques show it to be much slower than the diffusion-controlled rate of about 10^{10} l mole^{-1} sec^{-1}.

A more useful classification of acids is into carbon acids on the one hand, and acids in which the proton is bound to oxygen, nitrogen, fluorine or sulphur on the other. Acids of the first class usually react much more slowly than those of the second class, though there are some exceptions to this rule. The reasons for this difference in rates are not altogether clear, and two different types of explanation have been given. According to one view, the transfer of a proton from acids of the second class is facilitated by the formation of a hydrogen bond to the base, thus lowering the activation energy required. Such hydrogen bond formation is known to occur readily with O—H, N—H, and F—H, but is almost absent for C—H. The other type of explanation appeals to the change of electronic structure which often takes place when a carbon acid loses a proton. A hydrogen atom attached to carbon is not usually appreciably acidic unless the conjugate base is stabilized by the transfer of negative charge to a more electro-negative atom (usually oxygen or nitrogen), or by a spread of charge over many carbon atoms: both cases are illustrated in Table 10. Under these circumstances the reaction involves changes in the electronic structure and in the positions of nuclei, and the necessity for these rearrangements may retard the

† The explanation of the slow neutralization originally given by Hantzsch (and still sometimes quoted) was more complex, involving a slow change to the aci-form, e.g. $CH_3CH:NO.OH$, followed by a rapid neutralization by hydroxide ions. However, it is now believed that the aci-form could only be formed through the ion as an intermediate, rather than vice versa, and the neutralization is believed to consist of a single slow step.

reaction. It is probable that both of these factors are involved in the wide range of velocities observed.

In reactions involving carbon acids it is of interest to examine the rates of reaction of a given acid with a series of bases, or of a given base with a series of acids. One of the most extensive investigations of this kind was carried out by G. N. Lewis and G. T. Seaborg (1939), who examined the rate of neutralization of the anion of p-p'-p''-trinitrotriphenylmethane by a number of weak acids.† The reaction

Table 12

Rates of neutralization of the anion of p-p'-p''-*trinitrotriphenylmethane*

(Ethanolic solutions at $-60°C$. Velocity constant k in l mole^{-1} sec^{-1})

Acid	pK_a in water	k
Monochloroacetic	2·80	30
Furoic	3·15	10
α-Naphthoic	3·70	5·2
Lactic	3·85	3·1
Benzoic	4·15	5·0
Acetic	4·75	2·9
p-Nitrophenol	7·10	0·9
2,4-Dichlorophenol	7·55	0·8
Hydrocyanic	9·15	0·05
Boric	9·20	0·01
β-Naphthol	9·60	0·03
Phenol	9·85	0·01

was followed in ethanol solution at $-60°C$ by observing the colour change. The observed velocity constants (corrected for a slow reaction of the anion with the solvent) are given in Table 12, together with the pK-values of the acids in water.

It will be seen that there is a close parallelism between the velocity constant k and K_a, the dissociation constant of the acid. However, the range of values is much smaller for k than for K_a, and in fact the observed velocities can be represented quite well by an expression of the form $k = GK_a^{0·4}$, where G is a constant. Since the equilibrium

† The authors originally gave a more complex interpretation, but it would now be generally accepted that a simple acid-base reaction is involved.

F

constant, given by the ratio of the forward and reverse velocity constants, is directly proportional to K_a, it follows that the velocity constants for the reverse reaction (acid anion + p-p'-p''-trinitrotriphenylmethane) must follow a relation of the form $k' = G'K_a^{-0.6}$, where G' is another constant. This is in contrast with the behaviour already noted for the much faster reactions of oxygen or nitrogen acids with H_2O or OH^- (Table 11), for which the reaction velocity in one direction is diffusion-controlled, and hence independent of K_a, while in the reverse direction it is directly proportional to K_a or to $1/K_a$. We shall see shortly that relations involving K_a raised to a fractional power are common in acid-base catalysis.

The structural rearrangements which often accompany the interconversion of an acid and its conjugate base are of decisive importance for an understanding of *acid-base catalysis*. The catalytic power exerted by acids and bases in many reactions was recognized at an early date, and the study of acid-base catalysis played an important part in the development of reaction kinetics and the theory of electrolyte solutions. Initially little was known about the mechanism of catalysis, which was often attributed to some vague and mysterious influence, but it is now generally accepted that the first step in acid-base catalysis involves an *acid-base reaction between catalyst and substrate*. If this reaction involves a structural change the resulting form of the substrate may have new physical or chemical properties, leading to further reaction or other observable phenomena.

Thus for example the reaction:

$$CH_3NO_2 + CH_3COO^- \rightleftharpoons CH_2 : \overset{+}{N} \overset{\displaystyle O^-}{\underset{\displaystyle O^-}{\Big\langle}} + CH_3COOH$$

proceeds to a minute extent from left to right, and could not be detected by ordinary means. However, if in place of nitromethane we take an optically active nitroparaffin, $R_1R_2CH \cdot NO_2$, the anion $R_1R_2C : \overset{+}{N} \overset{\displaystyle O^-}{\underset{\displaystyle O^-}{\Big\langle}}$ is no longer asymmetric, and has an equal chance of reverting to either optical isomer: hence the rate of formation of this anion is equal to the *rate of racemization* of the nitroparaffin, and we should describe the racemization as being catalysed by acetate ions. Exactly the same position arises with an optically active ketone $R_1R_2CH.CO.R_3$, which is racemized by bases by conversion to the anion

$$R_1R_2C\!:\!C.R_3$$
$$\mid$$
$$O^-$$

Here again a structural change is involved.

Many other reactions depend upon the rate of formation of an anion from a weakly acidic substrate by the loss of a proton to a basic catalyst. The simplest is perhaps *deuterium exchange*: when the system is in a solvent containing both hydrogen and deuterium the anion may pick up either isotope when it reverts to the original substance. Catalysed deuterium exchange has been studied for many compounds, including ketones and nitroparaffins, and is often a convenient way of detecting very weak acidity: for example, acetylene exchanges deuterium with aqueous solutions containing a high concentration of hydroxide ions. Another frequent reaction of anions is *halogenation*, the classical example being the *acetome-iodine reaction* (A. Lapworth, 1904; H. M. Dawson, 1913–30). In this reaction the hydrogen atoms of the ketone are substituted by iodine, but the rate of reaction is independent of the iodine concentration, though it is affected catalytically by the presence of acids and bases in the solution. Moreover, if chlorine or bromine is used in place of iodine the rate still remains the same. These facts show that the process whose rate is being measured involves only the ketone and the catalysts, and not the halogen. In the presence of basic catalysts (e.g. OH^-, CH_3COO^-), the slow reaction is believed to be:

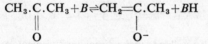

followed by very fast reaction of the anion with halogen.† In these reactions the halogen is often described as a *scavenger*, meaning a

† This interpretation differs slightly from that originally advanced by Lapworth, who believed that the enolic form of acetone was formed slowly and then reacted rapidly with halogen. We shall see shortly that this mechanism is still considered correct for many acid-catalysed halogenations.

reagent which removes a reactive species so rapidly that it does not have time to revert to its precursor.

The same type of zero-order halogenation occurs with a large number of compounds (e.g. ketones, esters, nitroparaffins, sulphones), and the anions formed can also undergo reaction with other organic molecules, for example in aldol and Claisen condensations. If the above reaction scheme is correct there should be correspondences between the rates of different processes: for example, under the same conditions of solvent, temperature, and catalyst an optically active ketone should have identical rates of racemization, iodination, bromination, and deuterium exchange. The equality of these rates has in fact been established in a number of instances (C. K. Ingold; P. D. Bartlett; O. Reitz, 1934–7).

The examples considered so far involve a reaction in one step between catalyst and substrate, but there are many reactions where two consecutive steps are involved. This is the case for the *catalysed interconversion of tautomers*, for example in keto-enol tautomerism. We do not now believe that a hydrogen atom can easily migrate directly from one part of a molecule to another, but rather that the process involves the addition and removal of protons. The following mechanisms are generally accepted for keto-enol tautomerisms, where A is any acid and B any base:

(32) *Acid catalysis*

$$\overset{(a)}{\underset{\diagdown}{\diagup}}CH.C:O+A \rightleftharpoons \overset{\diagdown}{\underset{\diagdown}{\diagup}}CH.\overset{+}{C}:OH+B \overset{(b)}{\rightleftharpoons} \overset{\diagdown}{\underset{\diagup}{}}C:C.OH+A$$

(33) *Basic catalysis*

$$\overset{\diagdown}{\underset{\diagup}{}}CH.C:O+B \rightleftharpoons \overset{\diagdown}{\underset{\diagup}{}}C:C.O^- +A \rightleftharpoons \overset{\diagdown}{\underset{\diagup}{}}C:C.OH+B$$

It is not often possible to make direct measurements on tautomeric interconversion, but under conditions of acid catalysis the rate of enol formation is equal to the rate of halogenation, racemization, or deuterium exchange. This is because the formation of the cation in (32) (unlike the anion in (33)) does not bring about any of these processes: on the other hand the enol is very reactive towards halogens

and similar reagents, and its formation from an optically active
ketone involves the loss of the asymmetric centre. When two consecutive proton-transfers are involved, as in the above scheme, one
of them will usually be much slower than the other, and will therefore determine the rate of the whole process. It is often difficult to
decide which steps are rate-determining, but in equations (32) and
(33) it is believed that steps (*a*) and (*d*) (involving no structural
change) are fast, while (*b*) and (*c*) (involving the shift of a double
bond) take place at a measurable speed.

Another important reaction which involves several steps is the
reversible *acid hydrolysis of esters*,

$$R'.COOR + H_2O \rightleftharpoons R'.COOH + ROH.†$$

A variety of mechanisms have been proposed, but many of them are
in fact equivalent, differing only in the extent to which the reaction is
dissected into steps. The following version is that given by J. N. E.
Day and C. K. Ingold (1941).

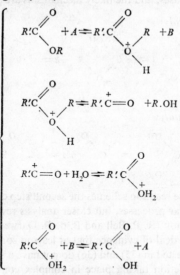

† The hydrolysis of esters can of course also be effected irreversibly by alkali,
but this reaction is a replacement of OR^- by OH^- rather than an acid-base
reaction, and will not be considered here.

Under different conditions different steps may become rate-determining.

We shall give one further example of the mechanism of a catalysed reaction, namely the *reversible addition of hydroxy-compounds to the carbonyl group*,

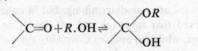

This type of reaction is involved in the reversible hydration of carbon dioxide and of aldehydes, the isotopic exchange of oxygen between ketones and water, the formation and decomposition of semi-acetals, the dimerization of α-hydroxy-aldehydes and ketones, the mutarotation of glucose, the exchange of alkyl groups between esters and alcohols, and the mutarotation of some α-keto-esters in presence of alcohols. Most of these processes are catalysed both by acids and by bases, and the likely mechanisms are as follows:

(35)   *Acid catalysis*

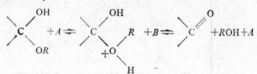

(36)   *Basic catalysis*

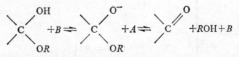

In each of these reaction schemes the second step can be divided into two bimolecular processes, but closer analysis reveals objections to any such division (R. P. Bell and B. de B. Darwent, 1950). There is in fact a good deal of evidence that at least in some cases the intermediates suggested in (35) and (36) do not have a separate existence, the whole reaction taking place in complex containing substrate, catalyst, and possibly solvent molecules in addition (M. Eigen, 1965; R. P. Bell, J. P. Millington and J. M. Pink, 1967).

Most of the quantitative laws governing the kinetics of catalysed reactions were established independently of any considerations about mechanism, and at an earlier date. One of the first points to be considered was the dependence of the reaction velocity upon catalyst concentration, and it was often found that for a given reaction in aqueous solution this could be represented by the equation:

$$v = v_0 + k_{H^+}[H^+] + k_{OH^-}[OH^-], \qquad (37)$$

v_0 is termed the *spontaneous rate*, and k_{H^+} and k_{OH^-} are the *catalytic constants* for hydrogen ion and hydroxide ion respectively. The values of v_0, k_{H^+}, and k_{OH^-} will of course depend upon the temperature and the nature of the reaction, and one or more of them may be zero for some reactions.

Equation (37) implies that the hydrogen and the hydroxide ions are the only effective catalysts, and this view met with much success in the early days of the ionic theory (*see* Tables 1 and 2, Chapter 1). However, some discrepancies gradually became evident, and subsequent progress depended to a large extent on a closer analysis of these. Some of them were *salt effects* of the kind already discussed in Chapter 4, and a failure to take these effects into account led to some misinterpretation. However, salt effects can usually be made small by carrying out experiments at low salt concentrations, or better still at constant ionic strength, and we shall not consider them further. A more important type of discrepancy led to the discovery of *general acid-base catalysis*. This term implies that catalysis may be exerted not only by hydrogen and hydroxide ions, but also by any species which are acids or bases in the sense of the Brönsted-Lowry definition discussed in Chapter 1. To take a specific example, if the catalysing solution is a buffer solution containing acetic acid and acetate ions, then the general expression for the velocity will be:

$$v = k_0[H_2O] + k_{H^+}[H^+] + k_{OH^-}[OH^-]$$
$$+ k_{HOAc}[HOAc] + k_{OAc^-}[OAc^-] \qquad (38)$$

in place of (37). The first term, representing the 'spontaneous' rate, has been written as $k_0[H_2O]$ in place of v_0, since it is now regarded as representing catalysis by water molecules in virtue of their acidic or basic properties.

The general nature of acid-base catalysis appears as a natural consequence of the definition of acids and bases, and the type of mechanism for catalysed reactions outlined above. As long as the hydrogen ion was believed to be a bare proton it was reasonable to believe that it could have unique catalytic properties (though no such claim could be reasonably made for the hydroxide ion) but as soon as it was accepted as the hydronium ion OH_3^+ other species would be expected to share its catalytic power. General acid catalysis was first established by H. M. Dawson and his collaborators for the acetone-iodine reaction (1913–30). General basic catalysis was found by J. N. Brönsted and K. J. Pedersen (1923) for the decomposition of nitramide, and somewhat later J. N. Brönsted and E. A. Guggen-heim simultaneously with T. M. Lowry and G. F. Smith (1927) showed that the mutarotation of glucose was subject to general catalysis both by acids and by bases. Both Brönsted and Lowry realized the bearing of these facts on the definition of acids and bases, and their observations on catalysis constituted part of the evidence in favour of the Brönsted-Lowry definition. Subsequently general catalysis by acids or bases (or both) has been established for a large number of reactions, and its presence or absence is often used to obtain information about reaction mechanisms.

As an example of the demonstation of general catalysis we shall consider some of the data for the decomposition of nitramide mentioned above. This substance decomposes in aqueous solution according to the equation $H_2N_2O_2 \rightarrow N_2O + H_2O$, and the reaction can be followed by observing the evolution of nitrous oxide. The reaction is strictly of the first order, and in solutions of strong acids the velocity constant is independent of hydrogen-ion concentration over a large range: this serves to demonstrate that $k_{H^+} = 0$ in equation (38), and to establish the value of k_0. Further information is obtained by measurements in buffer solutions, and Table 13 shows some of the data obtained in two different investigations for acetate buffers. It will be seen that the velocity bears no relation to the concentration of hydrogen ions, hydroxide ions, or acetic acid molecules, but is a linear function of the concentration of acetate ions. The last column shows velocities calculated from the expression $v = 38 \cdot 0 \times 10^{-5} + 0 \cdot 500[CH_3COO^-]$: the first term represents the velocity constant in solutions of strong acids, and $0 \cdot 500$ is the catalytic constant of the

acetate ion. Catalysis by hydroxide ions can be detected in more
alkaline solutions but is negligible at the pH of the acetate buffers
used. General basic catalysis occurs similarly in solutions containing uncharged bases: for example, in a buffer of aniline and its

Table 13

Decomposition of nitramide at 15°C.

v = first-order velocity constant, $\log_{10}$, minutes^{-1}

[CH$_3$COO$^-$]	[CH$_2$COOH]	[H$^+$]	10^5v obs.	calc.	Authors
0·00355	0·0146	8·2 × 10^{-5}	212	216	(2)
0·00397	0·0338	1·7 × 10^{-4}	234	237	(2)
0·00414	0·0162	7·8 × 10^{-5}	246	245	(1)
0·00683	0·0135	4·0 × 10^{-5}	382	380	(1)
0·00961	0·0169	3·5 × 10^{-5}	526	519	(2)
0·0102	0·0101	2·0 × 10^{-5}	551	548	(1)
0·0136	0·0067	9·8 × 10^{-6}	726	718	(1)
0·0158	0·0126	1·7 × 10^{-5}	800	818	(2)

(1). J. N. Brönsted and K. J. Pedersen (1923).
(2). E. C. Baughan and R. P. Bell (1937).

hydrochloride the most important catalytic species is the aniline
molecule. The mechanism of catalysis by a base B is probably:

$$NH_2NO_2 \rightleftharpoons NH{:}NO.OH$$
$$NH{:}NO.OH + B \rightarrow N{:}N.O + OH^- + A$$
$$A + OH^- \rightarrow B + H_2O$$

where the first and third reactions are very fast, and the second
is rate-determining. This provides an interesting example of an >NH
acid which reacts relatively slowly with bases because its loss of a
proton is accompanied by a drastic rearrangement of nuclei and
electrons.

In spite of the widespread occurrence of general acid-base catalysis,
there are a few reactions which appear to be *specifically catalysed by
hydrogen or hydroxide ions*: i.e. the reaction velocity is directly proportional to [H$^+$] or [OH$^-$], and independent of the presence in

solution of other acidic or basic species (including the water molecule). This does not necessarily mean that these reactions differ in principle from those exhibiting general catalysis, since the quantitative relation between the catalytic coefficients of different species may be such as to make H^+ or OH^- the predominant catalyst under ordinary experimental conditions. Reactions specifically catalysed by hydrogen ions in aqueous solution include the decomposition of ethyl diazoacetate

$$CHN_2.COOEt + H_2O \rightarrow CH_2OH.COOEt + N_2$$

and the hydrolysis of acetals,

$$CH_3CH(OR)_2 + H_2O \rightarrow CH_3CHO + 2ROH,$$

while the depolymerization of diacetone alcohol

$$(CH_3)_2C(OH).CH_2.CO.CH_3 \rightarrow 2(CH_3)_2CO$$

is specifically catalysed by hydroxide ions. Kinetic measurements on these reactions have sometimes been used for measuring concentrations of hydrogen or hydroxide ions, since they are free from complications due to catalysis by other species, and no acid is produced or consumed during reaction.†

We have so far considered only catalysis in aqueous solution, but much work has also been done in *non-aqueous solvents*. Some of the quantitative aspects of the results are not yet fully understood, but there are some general points of interest. This is particularly the case in *aprotic solvents* (e.g. hydrocarbons) where the solvent molecules can neither gain nor lose a proton, so that there are no analogues of the hydrogen and hydroxide ions. Similarly, there should be no 'spontaneous' reaction in these solvents, and it is in fact found that reactions such as keto-enol transformations can be arrested completely in hydrocarbon solvents provided that sufficient precautions are taken to exclude chance acidic or basic impurities. If an acid or a base (e.g. acetic acid, or ethylamine) is added to the solution, catalysis takes place, and since no reaction with the solvent is possible this catalysis can only be attributed to the molecules of

† It should be noted that the reaction velocity is more closely proportional to the concentration than to the activity of H^+ or OH^-. In acid solutions of high ionic strength the velocity is frequently proportional to the acidity function h_0 mentioned at the end of Chapter 4.

added substance. This is in contrast to the compound nature of catalysis in aqueous solution (cf. equation (38) above), and constitutes a direct demonstration of general catalysis by acids or bases. Such catalysis can even take place in the absence of any liquid phase, when both the catalyst and the substrate are sufficiently volatile to exist as gases: for example, both the depolymerization of paraldehyde and the dimerization of cyclopentadiene are catalysed by gaseous hydrogen chloride and similar substances (R. P. Bell and R. le G. Burnett, 1937–9; A. Wassermann, 1942–9). In most cases such catalysis takes place upon the wall of the reaction vessel, though the catalysed dimerization of cyclopentadiene is a homogeneous gas reaction.

By making quantitative kinetic measurements either in aqueous or in non-aqueous solutions it is possible to obtain data for the relative catalytic effects of a number of acidic or basic species for a given reaction, and it is often found that there is an approximate relation between the catalytic constant of a species and its strength as an acid or base. This was first established by Brönsted and Pedersen for the decomposition of nitramide, and the type of equation which they proposed is termed the *Brönsted relation*. For acid catalysis it has the form:

$$k_A = G_A K_A^{\alpha}, \tag{39}$$

where G_A and α are constants characteristic of the solvent, temperature, and reaction studied, α being positive and less than unity. k_A is the catalytic constant of a given acid, and K_A some measure of its acid strength, commonly its dissociation constant in water. An exactly analogous relation holds for catalysis by bases. Table 14 shows data for the acid-catalysed iodination of acetone, discussed qualitatively above. The agreement is fairly good here, since all the acids are of the same chemical type. Larger deviations occur if catalysts of widely varying structure and charge type are compared, but an equation such as (39) will usually predict the effectiveness of any catalyst to better than a power of ten. It is clear that equation (39) will break down if the velocity constants approach the diffusion-controlled limit, since the rate will then become independent of acid-base strength. This was originally pointed out by Brönsted and

Pedersen in 1923, and it has recently been stressed by Eigen (1963) that the most general plot of log k_A against log K_A must be a curve with limiting slopes of unity and zero for very slow and very fast reactions respectively. In practice the experimental data for a given reaction will normally cover only a small section of this curve, which is indistinguishable from a straight line with a slope intermediate between zero and unity. However, the curvature can often be detected if a very large range of velocities is covered, especially by using techniques for studying fast reactions.

Table 14

Acid catalysis in the iodination of acetone at 25°C.

Calculated values from $k_A = 7.9 \times 10^{-4} K_A^{0.62}$

Catalyst acid	K_A	$10^6 k_A$ obs.	$10^6 k_A$ calc.
Dichloroacetic	5.7×10^{-2}	220	270
$\alpha,\beta.$ Dibromopropionic	6.7×10^{-3}	63	54
Monochloroacetic	1.41×10^{-3}	34	32
Glycollic	1.54×10^{-4}	8.4	7.9
β-Chloropropionic	1.01×10^{-4}	5.9	6.2
Acetic	1.75×10^{-5}	2.4	2.2
Propionic	1.34×10^{-5}	1.7	1.8
Trimethylacetic	9.1×10^{-6}	1.9	1.5

We have already seen (Table 12) that a relation of the form of equation (39) holds for an acid-base reaction which can be studied directly, and it is reasonable that it should also apply to catalysed reactions. It would also be expected that, for a given catalyst, the reaction velocity would depend upon the acid-base strength of the substrate. This kind of relationship is more difficult to detect experimentally, because the substrate is commonly such a weak acid or base that direct measurements of its strength are not practicable. However, indirect support often comes from the effect of substituents: for example, in presence of bases, monochloroacetone halogenates about 1000 times as rapidly as acetone itself, corresponding to the well-known effect of chlorine substitution in increasing acidity (cf. Chapter 5). This kind of consideration can be used to estimate the acid-base strength of very weak substrates, and the value $pK = 19$, given in Table 10 for acetone, is based on such an argument. It should, however, be noted that such relationships hold only for substrates of closely similar structure: thus nitroethane, acetylacetone, and phenol are acids of closely similar dissociation constants, but under comparable conditions acetylacetone ionizes about 1000 times as fast as nitroethane, and phenol about 10^9 times as fast. Caution must therefore be exercised in deducing the strengths of acids or bases from the rates of their protolytic reactions.

GENERAL REFERENCES

J. N. Brönsted, 'Acid-base Catalysis', *Chem. Reviews*, 1928, **5**, 231.

R. P. Bell, *Acid-base Catalysis* (Oxford, 1941).

R. P. Bell, 'Acid-base Catalysis and Molecular Structure', *Advances in Catalysis* (New York, 1952).

R. P. Bell, *The Proton in Chemistry* (Ithaca, N.Y., and London, 1959), Chapters VIII and IX.

M. Eigen, *Angewandte Chem.*, 1963, **75**, 489.

Investigation of Rates and Mechanisms of Reactions, Vol. VIII, Part II, ed. S. L. Friess, E. S. Lewis, and A. Weissberger (New York and London, 1963).

E. F. Caldin, *Fast Reactions in Solution* (Blackwell, Oxford, 1964).

CHAPTER 7

Hydrogen Isotopes in Acid-base Reactions

In addition to the common form of hydrogen, of mass 1·008, there exist two other isotopes of this element, *deuterium* (symbol D) of mass approximately 2, and *tritium* (symbol T) of mass approximately 3. Naturally occurring hydrogen, whether as the element or in compounds, contains about 0·014% of deuterium, and the isotope of mass 1 is sometimes referred to as *protium* in order to distinguish it from the mixture: the isotope ^1H will however be designated hydrogen in this chapter. Various processes are available for increasing the proportion of deuterium, in particular the electrolysis of water, and both deuterium gas and a variety of deuterium compounds (e.g. D_2O, D_2SO_4, C_6D_6, CD_3COOD, etc.) are now available commercially. Tritium is a radioactive isotope which does not occur naturally to any appreciable extent, but is prepared artificially by nuclear reactions such as $^6_3Li + ^1_0n \rightarrow ^3_1T + ^4_2He$. Many compounds containing tritium can also be purchased, though only a small proportion of their hydrogen has been replaced by tritium: however, since the radioactivity of tritium provides a very sensitive and selective means of detecting it, much interesting chemistry can be done with such preparations. Most of this chapter will deal with observations on deuterium compounds, though many of its conclusions will also apply, *mutatis mutandis*, to tritium compounds.

Ever since the discovery of deuterium by H. C. Urey in 1931 an increasing amount of work has been done on the physical and chemical properties of deuterium compounds, and in particular on their acid-base reactions. Apart from their intrinsic interest, such investigations often throw light on the detailed mechanism of such reactions. Both equilibria and reaction rates have been studied, and for both of these there are considerable quantitative differences between the behaviour of deuterium compounds and that of the corresponding hydrogen compounds. Since it is common experience that two isotopes of the same element have almost identical properties, both in the free element and in compounds, we shall first enquire why it is that the isotopes of hydrogen are exceptional in this respect.

The potential energy curve representing the dissociation of XH, where X is an atom or group of atoms, will have the general form shown in Figure 7. (The same type of curve will apply to dissociation into $X+H$, X^-+H^+, or X^++H^-.) Since interatomic and intermolecular forces depend essentially on the number and distribution

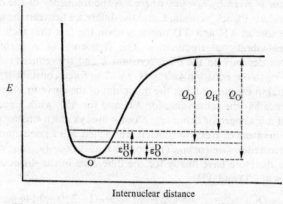

Fig. 7. Energy curves for the dissociation of XH and XD.

of electrons and the charges on the nuclei, and only to a very minor extent on the nuclear masses, the curve for the dissociation of XD will coincide with that for XH. On this basis we might expect identical dissociation energies for the two species XH and XD, represented by the distance Q_0 in Figure 7, and hence also equal equilibrium dissociation constants. Actually there are considerable differences between these quantities for the two isotopes, arising from differences in *zero-point energies*. It is a fundamental consequence of the quantum theory that a system capable of vibration can never exist in the motionless state represented by the point O in Figure 7: however low the temperature it will always retain a residual (or zero-point) energy represented in the figure by the levels

ε_0^H and ε_0^D. This follows qualitatively from the *uncertainty principle*, according to which it is impossible to specify simultaneously the exact position and the exact momentum of any system. The motionless state 0 is inconsistent with this principle, since it corresponds to zero momentum and an exactly defined internuclear distance for XH or XD; on the other hand, the states of energy ε_0^H and ε_0^D are permissible since they allow a range of values both for the internuclear distance and for the momentum.

A quantitative treatment shows that the zero-point energy of an oscillator is given by $\varepsilon_0 = \frac{1}{2}h\nu$, where ν is the frequency of the oscillator and h is Planck's constant, and the difference between the levels for the species XH and XD depends upon the fact that they have different vibrational frequencies. The frequency of a harmonic oscillator depends on the force constant k and the reduced mass μ according to the equation $4\pi^2\nu^2 = (k/\mu)^{\frac{1}{2}}$. The force constant is proportional to the curvature at the minimum of the curve in Figure 7, and hence has the same value for XH and for XD, while since the mass of a hydrogen or deuterium atom is always much smaller than that of the atom or group to which it is attached, it is a good approximation to write $\mu \simeq m_H$ or m_D. This implies that $\nu_H/\nu_D = (\mu_D/\mu_H)^{\frac{1}{2}} \simeq 2^{\frac{1}{2}}$, and we therefore have finally for the difference in the dissociation energies of XD and XH

$$Q_D - Q_H = \varepsilon_0^H - \varepsilon_0^D = \frac{1}{2}h(\nu_H - \nu_D) = \frac{1}{2}h\nu_H(1 - 2^{-\frac{1}{2}}) = 0.146\ h\nu. \quad (40)$$

Table 15 gives values of ν_H and $\varepsilon_0^H - \varepsilon_0^D$ for typical C—H, N—H, and O—H bonds. Although the energy differences amount to only a small fraction of the dissociation energy, which is about 100,000 cal/mole (so that the height of the levels ε_0^H and ε_0^D in Figure 7 is greatly exaggerated), they are nevertheless large enough to produce considerable differences in the dissociation equilibrium constants for the two isotopes. If we can identify the energy difference with a free energy difference the relation is $K_H/K_D = \exp\{(\varepsilon_0^H - \varepsilon_0^D)/kT\}$, and values of this quantity at 25°C are given in the last column of Table 15.

Isotopes of other elements will of course lead to differing zero-point energies, and the above considerations will apply equally well in principle. However, there are three reasons why all elements other than hydrogen will give rise to very much smaller isotope effects: (a) the vibrational frequencies are lower, (b) the ratio of the isotopic masses is much closer to unity, and (c) the reduced mass for a vibration is now less affected by a change in the mass of one of the vibrating groups. For example, consider the dissociation of the two bonds ^{12}C—^{12}C and ^{12}C—^{13}C, both having a frequency close to 900 cm^{-1}. The difference in zero-point energies now amounts to only 0·010 $h\nu$, i.e. 26 cals/mole, and this corresponds to a difference of only 4% in the equilibrium constants for dissociation. The isotopes of hydrogen thus constitute a very special case.

Table 15

Zero-point energies for stretching vibrations in hydrogen and deuterium compounds

Bond	ν_H (cm$^-$)	$\varepsilon_0{}^H - \varepsilon_0{}^D$ (cal/mole)	exp $\{(\varepsilon_0{}^H - \varepsilon_0{}^D)kT\}$ at 25°C
C—H	2,800	1,150	6·9
N—H	3,100	1,270	8·5
O—H	3,300	1,400	10·6

As we have seen in Chapter 1, acid-base reactions in solution never lead to the splitting off of a free proton, but involve a reaction between the two acid-base pairs, one of which may be derived from the solvent. A more realistic scheme for considering isotope effects is therefore represented by equation (41).

$$XH + Y^- \overset{k_H}{\rightleftharpoons} X^- + HY, \text{ equilibrium constant } K_H$$
$$XD + Y^- \overset{k_D}{\rightleftharpoons} X^- + DY, \text{ equilibrium constant } K_D \tag{41}$$

The ratio of the equilibrium constants, K_H/K_D, also represents the equilibrium constant for the exchange reaction $XH + DY \rightleftharpoons XD + HY$. The energy profile for reactions (41) is shown schematically by Figure 8, where the ordinate no longer represents any single internuclear distance, but may be regarded for our purpose as an unspecified quantity (sometimes called the 'reaction co-ordinate') which increases continuously as the proton is transferred from X to Y. The presence of a maximum in the curve indicates that the reaction has

G

an appreciable activation energy in both directions, though this feature of the diagram is irrelevant when (as in the present section) we are considering equilibria. Both XH and HY have zero-point energies, which will have different values if the hydrogen is replaced by deuterium, as shown on the diagram. The energy changes for the two reactions in (41) are indicated in Figure 8 by Q_H and Q_D, and it is clear that these two quantities will not be equal unless the frequencies (and therefore the zero-point energies) of XH and XD happen to

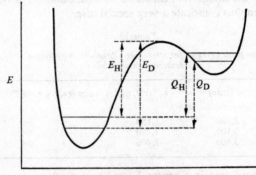

Reaction co-ordinate

Fig. 8. Energy curves for the reactions XH(XD)$+ Y^- \rightarrow X^- +$ HY(DY).

coincide with those of HY and DY respectively. The actual difference will be given by

$$Q_D - Q_H = \varepsilon_0(XH) - \varepsilon_0(XD) - \{\varepsilon_0(HY) - \varepsilon_0(DY)\}. \qquad (42)$$

However, since the differences in zero-point energy will always cancel out to some extent, the values of $Q_D - Q_H$ will be smaller than the differences $\varepsilon_0^H - \varepsilon_0^D$ given in Table 15, and the ratio K_H/K_D will deviate less from unity for proton-transfer reactions such as (41) than for a simple dissociation.

The zero-point energy differences are often the major factor in determining isotope effects on equilibria, but it may be necessary also to allow for the different spacing of higher vibrational levels: further, since the equilibrium constants are related to free energies rather than total energies, it is also necessary in a complete treatment to take into account the effect of isotopic mass on rotational and vibrational entropies. The methods for doing this are contained in the standard formulae of statistical mechanics, and will not be discussed here. It turns out (H. C. Urey, 1947; J. Bigeleisen and M. G. Mayer, 1947) that the isotope effect can be expressed solely in terms of the masses of the isotopic atoms involved and the vibrational frequencies of the reacting species. For fairly simple molecules the latter can be obtained from observed spectra, and successful predictions have been made for a number of gas reactions. Examples are the exchange reaction $H_2 + 2DI \rightleftharpoons D_2 + 2HI$ (the equilibrium constant for which is obtained experimentally as the ratio of the constants for the two equilibria $H_2 + I_2 \rightleftharpoons 2HI$ and $D_2 + I_2 \rightleftharpoons 2HD$) and the reaction $H_2 + HDO \rightleftharpoons HD + H_2O$, which is important in the industrial concentration of deuterium.

It might be thought that the same principles could be applied directly to the dissociation constants of two acids XH and XD, defined in terms of the reactions $XH + S \rightleftharpoons X^- + SH^+$ and $XD + S \rightleftharpoons X^- + SD^+$, where S is the solvent. However, this is not possible in most common solvents, notably water, which contain hydrogen atoms which exchange rapidly with the acids XH or XD. Thus it is not possible to compare the dissociation constants of CH_3CO_2H and CH_3CO_2D either in H_2O or D_2O, since CH_3CO_2H is converted almost completely to CH_3CO_2D in D_2O, and CH_3CO_2D to CH_3CO_2H in H_2O. The direct comparison can only be made in an aprotic solvent such as a hydrocarbon, with the addition of a base which contains no exchangeable hydrogen, for example a tertiary amine. A few measurements of this kind have been made (R. P. Bell and J. E. Crooks, 1962) and the results are in accordance with theoretical predictions, though it is necessary to take into account whether or not there is hydrogen-bonding in the ion-pair formed by reactions such as $XH + NR_3 \rightleftharpoons X^- \overset{+}{H} NR_3$. The observed effects are small; for example, for the reaction between dinitrophenol and pyridine in chlorobenzene solution $K_H/K_D = 1 \cdot 4$.

Almost all the available information on hydrogen isotope effects in acid-base equilibria consists of a comparison of the dissociation constant of XH in H_2O with that of XD in D_2O. The reactions concerned are therefore

$$XH + H_2O \rightleftharpoons X^- + H_3O^+ \text{ in } H_2O$$
$$XD + D_2O \rightleftharpoons X^- + D_3O^+ \text{ in } D_2O \tag{43}$$

and the constants can be denoted by K_{H_2O} and K_{D_2O}. Many measurements of this kind have been made, and a selection of the results is given in Table 16.

Table 16

Dissociation constants of acids in H_2O and D_2O at 25°C.

Acid	pK in H_2O	K_{H_2O}/K_{D_2O}
Picric	0·30	2·8
Cyanoacetic	2·50	2·8
Chloroacetic	2·87	3·1
4-Chloro–2,6-dinitropheno	2·97	3·1
2,6-Dinitrophenol	3·71	3·1
Formic	3·75	2·9
2,4-Dinitrophenol	4·09	3·3
Benzoic	4·21	3·1
Acetic	4·75	3·3
Trimethylacetic	5·03	3·1
2,5-Dinitropheno	5·22	3·4
p-Nitrophenol	7·24	3·7
o-Nitrophenol	7·25	3·7
Hydroquinone	10·58	4·2
2,2′,2″-Trifluoroethanol	12·37	4·5
2-Chloroethanol	14·31	5·0
Water	15·74	6·5

It will be seen that K_{H_2O} is always considerably greater than K_{D_2O}. Because of the change of solvent, the quantitative interpretation of this difference is more complicated than for equilibria studied in the same solvent. It can, however, be accounted for qualitatively by the fact that the stretching vibrations of the ion H_3O^+ have a considerably lower frequency (2900 cm^{-1}) than the corresponding vibrations in the water molecule (3400 cm^{-1}) or in the OH—groups of most organic compounds (average 3300 cm^{-1}). This means that the corresponding zero-point energy will be lower on the right-hand side of equation (43) than on the left-hand side, and this difference will be greater in H_2O than in D_2O: hence the degree of dissociation should be greater in the former. If the difference in the changes in zero-point energy is interpreted as a difference in free energies, this would predict a constant value of 3·8 for K_{H_2O}/K_{D_2O}, which is close to the average observed value. However, this quantitative agreement is probably coincidental, and in particular it does not account for the trend of K_{H_2O}/K_{D_2O} to higher values as the acid becomes weaker: this can be observed in Table 16, especially for the values relating to phenols, alcohols and water. The latter variation was at first attributed to an increase in the O—H frequency of the acid with increasing pK, but examination shows that this effect is a very small one. A complete interpretation of the problem must certainly take into account the hydrogen-bonding to the solvent of the species H_3O^+, XH and X^- (or their deuterium analogues), which for XH and X^- will involve frequencies which are dependent on the acid strength of XH (C. A. Bunton and V. J. Shiner, 1961). It is also probable that both bending and stretching frequencies must be considered in a complete treatment.

Turning now to the kinetics of acid-base reactions (including acid-base catalysis), it should be mentioned first that even on a classical basis (i.e. without invoking the quantum theory) there should be some differences between the reaction rates of hydrogen and deuterium compounds. This is because both collision frequencies and vibration frequencies are dependent on mass, and both of these will often be involved in determining reaction rates. However, it is easily shown that for two isotopes A and B the maximum kinetic isotope which can be explained on this basis is $k_A/k_B = (m_B/m_A)^{\frac{1}{2}}$, i.e. $\sqrt{2}$ for hydrogen and deuterium. In fact very much larger effects are observed in practice, and again the cause lies in the different zero-point energies of the links involving hydrogen and deuterium. This is illustrated by the activation energies E_H and E_D marked in Figure 8. In the transition state there is no vibrational energy in the direction of the reaction co-ordinate, since the energy of the system is at a maximum rather than a minimum, and motion along this co-ordinate has the nature of a translation rather than a vibration. Because of the consequent absence of zero-point energy, the energy of the transition state will be the same for deuteron and proton transfer, and the difference of activation energies for reaction (41) from left to right is simply given by

$$E_D - E_H = \varepsilon_0(XH) - \varepsilon_0(XD) \qquad (43)$$

with an analogous expression involving HY and DY for reaction in the reverse direction. Since the reaction velocity is exponentially related to the activation energy, the values in the right-hand column of Table 15 should in fact represent the ratio of the velocity constants, k_H/k_D, for the forward reaction.

A large proportion of the experimental evidence on kinetic hydrogen isotope effects in acid-base reactions relates to the ionization of C—H bonds, and much of it has been obtained by the indirect methods described in Chapter 6 such as measurements of rates of racemization or isotope exchange, or the use of halogens as scavengers to react with the carbanions formed. Many of the reactions thus studied would usually be described as being catalysed by acids or bases. Since the ionization of C—H bonds is a relatively slow process, exchange with hydroxylic solvents is also slow, and it is possible to study both C—H and C—D compounds in a common solvent such as H_2O, thus avoiding complications caused by a change of solvent from H_2O to D_2O.

A small selection of the available data is given in Table 17. The deuterium rates refer to compounds in which all the ionizing hydrogens are replaced by deuterium: for example, for acetone the comparison is between CH_3COCH_3 and CD_3COCD_3.

Table 17 shows that effects of the expected order of magnitude are in fact observed and the presence or absence of a considerable hydrogen isotope effect is frequently used for deciding details about the mechanism of reactions. Since only a semi-quantitative result is required, a direct comparison between experiments in H_2O and D_2O is sometimes sufficient. For example, the hydrolysis of ethyl vinyl ether, catalysed by hydrogen ions, is believed to involve the following steps:

$$CH_2:CH.O.C_2H_5 + H_3O^+ \rightleftharpoons CH_3.CH:O^+.C_2H_5 + H_2O$$

$$CH_3.CH:O^+.C_2H_5 + 2H_2O \rightarrow CH_3CHO + C_2H_5OH + H_3O^+$$

where the second step may be split up into two stages, involving the formation of a free carbonium ion. It is found that this reaction takes place 3·2 times more slowly in D_2O than in H_2O, and this suggests strongly that the first step, which is a proton-transfer to carbon, is the rate-determining one, followed by a rapid second step.

Table 17

Kinetic isotope effects in the ionization of C—H bonds in water at 25°C.

Acid	Base	k_H/k_D
$CHFCl_2$	OH	1·5
CHFClBr	OH^-	1·7
CH_3NO_2	H_2O	3·8
CH_3NO_2	$CH_2ClCO_2^-$	4·3
CH_3NO_2	$CH_3CO_2^-$	6·5
CH_3NO_2	OH^-	10·3
CH_3COCH_3	OH^-	9·8
$(CH_3CO)_2CH_2$	H_2O	4·5
$SO_3^-CH_2COCH_3$	H_2O	2·4
$SO_3^-CH_2COCH_3$	CH_2ClCO_2	2·9
$SO_3^-CH_2COCH_3$	$CH_3CO_2^-$	3·9
$SO_3^-CH_2COCH_3$	$Me_3C.CO_2$	4·4
$SO_3^-CH_2COCH_3$	2,6-lutidine	7·2
$SO_3^-CH_2COCH_3$	OH^-	9·4

By contrast, the acid-catalysed hydrolysis of ethyl diazoacetate, which also involves two steps

$$\overset{-}{N}:\overset{+}{N}:CH.CO_2Et + H_3O^+ \rightleftharpoons \overset{+}{N}:N.CH_2CO_2Et + H_2O$$
$$\overset{+}{N}:N.CH_2CO_2Et + 2H_2O \rightarrow N_2 + CH_2OHCO_2Et,$$

is found to be 2·7 times *faster* in D_2O than in H_2O. This excludes the initial proton-transfer as a rate-determining process: the inverse isotope effect relates to the initial equilibrium (which, in reverse, represents the dissociation of a weak acid in H_2O or D_2O), followed by the second rate-determining stage having little or no isotope effect. This distinction between the two reactions is confirmed by the observation that when they are carried out in deuterium oxide the unchanged vinyl ethyl ether contains no deuterium, while the unchanged diazo-compound does: this means that the reverse of the first step takes place readily for the diazo-compound, but to a negligible extent for the vinyl ether.

Another class of reaction in which hydrogen isotope effects have been used to obtain detailed information about reaction mechanisms is *electrophilic aromatic substitution*. For benzene itself the reaction sequence is

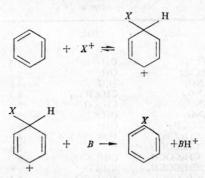

where X^+ is the electrophilic reagent (e.g. NO_2^+) and B is a base, commonly the solvent. The problem is again to decide whether the formation of the intermediate cation (often called the Wheland intermediate) is rate-determining, or whether it is in equilibrium with the reactants and then undergoes a rate-determining proton-transfer reaction with base. In many instances of nitration it has been shown (first by L. Melander, 1949) that deuteriation or tritiation of the aromatic substance has little effect on the velocity, so that the first step must be rate-determining.†

The opposite behaviour is found in other types of reaction, for example in some azo-coupling reactions, where the electrophilic reagent is ArN_2^+ (H. Zollinger, 1955). Thus the reaction of 4-chloro-diazobenzene with 2-naphthol-6,8-disulphonic acid shows a large isotope effect ($k_H/k_D = 6\cdot6$), and the kinetic importance of the second step is confirmed by the fact that the reaction is accelerated

† This was first established by competition methods, in which the tritium content of an aromatic substance was found to be unchanged on sulphonation or nitration.

by the addition of a base such as pyridine. Moreover, at high pyridine concentrations the isotope effect becomes smaller: this is because the second step is speeded up so much that it is no longer completely rate-determining.

Considerable deuterium isotope effects are also observed in the iodination of phenols, and here another type of correlation with the reaction conditions is observed. When molecular iodine is the main iodinating agent the reaction sequence is

$$ArH + I_2 \rightleftharpoons ArHI^+ + I^-$$
$$ArHI^+ + B \rightarrow ArI + BH^+$$

As the concentration of iodide ions is decreased, the reverse of the first reaction is retarded: this again prevents the second reaction from being fully rate-determining, and a decreased isotope effect is therefore observed at low iodide concentrations (E. Grovenstein and N. S. Aprahamian, 1962).

It would be easy to multiply the above examples, and the semi-quantitative use of hydrogen isotope effects for detecting rate-determining proton-transfers is now a normal procedure in mechanistic organic and inorganic chemistry. However, the quantitative interpretation of results such as those in Table 17 is still incompletely understood. Although all the reactions listed there involve the ionization of C—H bonds, the values of k_H/k_D vary considerably, and many of them are appreciably lower than the value of 7 predicted from the zero-point energy of C—H bond stretching (cf. Table 15). In particular, it can be seen from Figure 8 and equation (43) that for a given acid $XH(XD)$, the difference of activation energies $E_D - E_H$, and hence the magnitude of the isotope effect, should be independent of the nature of the base Y which abstracts the proton or deuteron. The data for nitromethane and the acetone-sulphonate ion given in Table 17 show that this is far from the case, the value of k_H/k_D increasing regularly with the strength of the base. These variations are probably due to the fact that some of the initial zero-point energy is retained in the transition state to an extent depending upon its exact structure. This problem will not be pursued further here, but it may be mentioned that the quantitative study of kinetic isotope effects promises to be one of the most

H

fruitful methods for obtaining information about the structure of transition states, and hence about the detailed mechanism of chemical reactions.

Little experimental evidence is available about hydrogen isotope effects in the much faster transfers of protons between oxygen or nitrogen atoms. When these are exothermic (cf. Table 13), no isotope effect would be normally anticipated, since the reaction velocity depends on a rate of diffusion rather than on a chemical activation energy. A case of particular interest is the *mobility of the hydrogen ion* in water, the high value of which is attributed to proton transfer reactions of the type

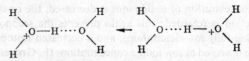

The mobilities of excess protons and deuterons in solid H_2O and D_2O have been measured (Eigen, 1963), and show a 'normal' isotope effect of $l_H/l_D \simeq 7$. On the other hand, in liquid water the mobility of protons is only about $1 \cdot 4$ times as great as that of deuterons: this is because the rate-determining step is now the rotation of the water molecules into the correct orientation for proton-transfer, which then takes place rapidly. The factor of $1 \cdot 4$ (approximately $2^{\frac{1}{2}}$) arises from the difference in the moments of inertia of H_2O and D_2O.

So far we have considered only *primary isotope effects*, in which bonds to hydrogen or deuterium are formed or broken during reaction. A large amount of work has also been done on *secondary isotope effects*, in which the binding of the isotopic atoms is un-affected by the reaction, at least in its conventional formulation. These effects are smaller than the primary ones, but their investigation may throw light on the details of binding in the species concerned, including transition states. An example of a secondary isotope effect on an equilibrium constant is the comparison of the two reactions

$$HCO_2H + H_2O \rightleftharpoons HCO_2^- + H_3O^+$$
$$DCO_2H + H_2O \rightleftharpoons DCO_2^- + H_3O^+.$$

Since the hydrogen attached to carbon does not exchange with water at an appreciable rate, both equilibria can be studied in H_2O, and it is found that $K(HCO_2H)$ is about 11% greater than $K(DCO_2H)$ (R. P. Bell and W. B. T. Miller, 1963). Although the H—C and D—C bonds are written formally in the same way in the acid and its anion, the observed isotope effect shows that they must in fact be somewhat weakened on ionization of the carboxyl group. This change is reflected in the infra-red and Raman frequencies of formic acid and the formate ion, and in fact a theoretical calculation on the basis of these frequencies successfully predicts the observed isotope effect (R. P. Bell and J. E. Crooks, 1962). Similar secondary equilibrium effects are observed in comparisons of the acid strengths of pairs such as CH_3CO_2H and CD_3CO_2H, or $CH_3NH_3^+$ and $CD_3NH_3^+$. As an example of a secondary kinetic isotope effect in a proton-transfer reaction we may quote the investigation by W. D. Emmons and M. F. Hawthorne (1956) of the rates of ionization of the two ketones

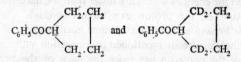

studied by measuring their rates of bromination. Although the same proton (underlined above) is being removed in each case, the replacement of hydrogen by deuterium in the adjacent methylene groups decreases the rate of ionization by about 20%. This again shows that the bonding in these methylene groups must be somewhat weakened in the transition state, though in this case no quantitative comparison with spectral frequencies can be made.

Finally, brief mention will be made of an interesting theoretical concept which has often been invoked to explain some features of kinetic hydrogen isotope effects, namely the *non-classical behaviour of light particles*. According to the quantum theory, all particles behave in part like waves, with a characteristic wavelength given by $\lambda = h/mv$, where mv is the momentum of the particle. On account of their low mass, electrons have wavelengths of the same order of magnitude as molecular dimensions, and in fact the whole of the quantum theory of atomic and molecular binding is closely related to this fact. Most nuclei are so heavy that their wavelengths are much smaller than molecular dimensions, so that their motion in chemical reactions can be treated to a good approximation by the laws of classical mechanics. In this respect the proton falls into a special position, since protons with thermal energies at room temperatures have a wavelength of about 0·8 Å, which is not much smaller than the distances through which protons or hydrogen atoms move in chemical reactions. Soon after the birth of modern quantum theory a number of authors (D. G. Bourgin, 1929; R. N. Langer, 1929; S. Roginsky and L. Rosenkewitsch, 1930; E. Wigner, 1932; R. P. Bell, 1933) suggested that the non-classical behaviour of the proton might be of importance in chemical kinetics, but until quite recently there was no clear experimental evidence to confirm this.

The consequences of the wave-nature of the proton are usually expressed in terms of the *tunnel effect*. This implies that, in a reaction such as that illustrated in Figure 8, there is a finite probability of reaction even for energies which are lower than the maximum in the energy curve: i.e., the proton can 'tunnel through' the energy barrier. This behaviour is in fact another manifestation of the un-certainty principle, already mentioned in connection with zero-point energy. Since the deuteron has twice the mass of the proton, its wavelength is smaller, and it behaves more like a classical particle; in particular, it is less susceptible to the tunnel effect. There may thus be a contribution to kinetic hydrogen isotope effects which is additional to that attributable to differences of zero-point energy, and abnormally high values of k_H/k_D might be anticipated, especi-ally at low temperatures.

It is noticeable that several of the values in Table 17 are consider-ably greater than the expected figure of 7, and although there may be other explanations of this increase, there are a few examples of much larger effects which can hardly be explained without invoking the tunnel effect; for example, the abstraction of a proton from 2-nitro-propane by the base 2,6-lutidine has $k_H/k_D = 20$–24 in two different solvents (E. S. Lewis and L. Funderburk, 1964; R. P. Bell and D. M. Goodall, 1966). The theory of the tunnel effect also predicts other anomalies, such as large differences between E_D and E_H, and devia-tions from the Arrhenius equation at low temperatures, both of which have been observed in some instances. It thus seems likely that the non-classical behaviour of the proton is a factor which will have to be taken into account in any full understanding of the kinetics of acid-base reactions in general, and of kinetic isotope effects in particular, though it is not yet clear how drastic or how widespread its consequences will be.

GENERAL REFERENCES

R. P. Bell, *The Proton in Chemistry* (Ithaca, N.Y., and London, 1959), Ch. XI.

L. Melander, *Isotope Effects on Reaction Rates* (New York, 1960).

H. Zollinger, 'Hydrogen Isotope Effects in Aromatic Substitution Reactions' (*Advances in Physical Organic Chemistry*, **2**, 1964).

CHAPTER 8

Alternative Uses of the Terms Acid and Base

We have so far used the terms acid and base in the sense of the Brönsted-Lowry definition, i.e. as proton-donors and proton-acceptors respectively. It is now common practice, especially in the United States, to use these terms in a different sense. This usage is commonly described as an 'extension' of the acid-base concept, but we shall see that it does in fact involve using the term acid for an essentially different group of substances. Such questions of nomenclature are largely a matter of taste, and in this chapter we shall give some reasons for restricting our considerations to proton-acids.

The alternative usage is mainly due to G. N. Lewis (1923–42), who based his views partly on theoretical and partly on experimental considerations. On the theoretical side he classifies as acid-base reactions those in which an unshared electron-pair in the base molecule is accepted by the acid molecule with the formation of a covalent link. This classification includes as bases the same species as the Brönsted-Lowry definition, since a molecule or ion which will add on a proton does so in virtue of an unshared pair of electrons, and will also combine with other electron-acceptors. On the other hand, typical Lewis acids are species with an outer electron shell capable of further expansion, e.g. BF_3, SO_3, Ag^+, none of which are acids in the Brönsted-Lowry sense. Moreover, classical acid-base reactions as usually written (e.g. $NH_3 + CH_3COOH \rightleftharpoons NH_4^+ + CH_3COO^-$) do not involve any donation or acceptance of electron-pairs, and in fact the acids of the older definitions (HCl, H_2SO_4, CH_3COOH, etc.) can only be included in the Lewis scheme by somewhat indirect means. Lewis supposes that the reaction between an acid XH and a base B is initiated by the formation of a hydrogen bond, XH$\cdots B$, in which the hydrogen accepts extra electrons from the base, thus justifying the inclusion of XH as an electron acceptor. However, the hydrogen bond bears little resemblance to an ordinary electron-pair bond, and for this reason 'ordinary' acids such as HCl, H_2SO_4, etc., are sometimes referred to as secondary acids by Lewis and his school. The electron acceptor molecules do in fact differ considerably from the Brönsted–Lowry acids from a theoretical point of view, and it seems desirable to distinguish them by different names. In the remainder of this chapter we shall refer to them as *Lewis acids* and *proton acids* respectively.

From the experimental point of view Lewis classes as acids all substances which exhibit 'typical' properties such as neutralization of bases, replacement, action on indicators, and catalysis, irrespective of their chemical nature or exact mode of action. Superficially this view appears to correlate a wide range of phenomena in the qualitative sense. For example, solutions of BF_3 or SO_3 in inert solvents bring about colour changes in indicators very similar to those produced by HCl, and these changes are reversed by adding bases, so that a titration can be carried out. Similarly, the same substances catalyse a large number of organic reactions, a few of which are catalysed by proton acids. The substances classified experimentally as Lewis acids are usually electron acceptors, but this is not always the case: for example, CO_2 and N_2O_5 contain completed octets and on ordinary valency theory cannot accept more electrons.

The value of this experimental generalization decreases somewhat when the individual cases are considered more closely. In the first place, the acidic properties of Lewis acids are sometimes due to their action on a hydrogen compound present in the system, thus releasing a proton. For example, it has been shown (M. Polanyi and co-workers, 1947) that Friedel-Crafts catalysts such as BF_3, $AlCl_3$, $TiCl_4$, etc., will not bring about the polymerization of iso-butene when both components are rigorously purified. The polymerization commonly observed in such systems is due to traces of a proton acid (e.g. H_2O, HCl, butyl alcohol), the initial reaction being, for example:

$$BF_3 + H_2O + CH_2{=}CHR \rightarrow [BF_3OH]^- + \overset{|}{\underset{+}{CH_2}}{-}CH_2R$$

the ion-radical then reacting with more olefin. Such co-operation between a Lewis acid and a proton acid is a fairly common phenomenon: for example, a mixture of HF and BF_3 is a very powerful reagent for transferring a proton to very weak bases, such as olefins and aromatic hydrocarbons, the ion BF_4^- being formed in the process.†

There are of course some reactions in which a Lewis acid alone can act as a catalyst, but it is only rarely that these are found to be catalysed by proton acids also. For example, the catalytic effect of $AlCl_3$ often depends upon a reaction of the type $RCl + AlCl_3 \rightarrow R^+ + AlCl_4^-$, R^+ being the reactive entity: proton acids do not act as catalysts in this type of reaction. One of the few reactions for which quantitative data are available on the catalytic effect of both Lewis acids and proton acids is the depolymerization of paraldehyde (R. P. Bell and B. Skinner, 1952) according to the equation:

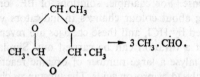

This reaction is peculiar in that it demands only a rearrangement of electrons, and not the migration of atoms or groups: it can in fact (unlike most catalysed reactions) take place at a reproducible rate in the absence of any catalyst. The function of both types of catalyst is to promote the electronic rearrangement by withdrawing electrons from the oxygen atom, either by forming

$$\diagdown OH^+, \text{ or (for example) } \diagdown \overset{+}{O} \!-\! \overset{-}{B}F_3.$$

Such similarity in the action of the two types of acid is not possible in the commoner class of reactions involving the migration of a proton.

The same kind of similarity is involved in the action of the two classes of acids upon indicators. The action of any proton acid on a basic indicator produces (by addition of a proton) the same species,

---

† This action is sometimes attributed to the intermediate formation of the very strong proton acid HBF_4. However, there is no detectable interaction between HF and BF_3 in the absence of a proton acceptor, and there seems to be no advantage in introducing the hypothetical HBF_4, to which no reasonable valency structure can be assigned.

which has its characteristic absorption spectrum. The action of a Lewis acid on the same indicator will produce a species which is different for each acid: however, the addition of a proton and of any Lewis acid will produce the same kind of electron displacements in the indicator molecule, so that the resulting changes of absorption spectra will be very similar, and may be indistinguishable. The two classes of acids have a similar function in this respect, but they are not identical.

The chief justification for a separate treatment of proton acids lies in the quantitative relations which they obey. As we have seen, a single constant specifying the strength of an acid-base pair is sufficient to predict accurately the position of its equilibrium with any other acid-base pair in the same solvent: moreover, the same constant will give an approximate estimate of equilibria in other solvents, and of the velocity of reactions involving the acid-base pair. These quantitative relationships represent the most valuable contributions of classical electrolyte theory and of the Brönsted–Lowry acid-base concept. Although less quantitative work has yet been done with Lewis acids, it is already clear that no such simple relations hold, even in a qualitative sense. G. N. Lewis himself has stated that 'the relative strengths of acids and bases depend not only upon the chosen solvent, but also upon the particular acid used for reference'. For example, in the classical sense ammonia is a much weaker base than hydroxide ion, but when referred to the Lewis acid Ag^+ the order of strengths is reversed, since AgOH is completely dissociated, while $Ag(NH_3)_2^+$ is a stable complex. This absence of any simple system of acid-base strengths is a high price to pay for an increased descriptive scope.

However, a recent development in the classification of Lewis acids and bases has at least introduced a guiding qualitative principle for deciding whether or not they will react readily with one another. This is the concept of *soft and hard acids and bases*, first introduced by R. G. Pearson (1963). The aim is to divide both acids and bases into two categories, soft and hard, in such a way that soft acids prefer to combine with soft bases and hard acids with hard bases. This idea of course explicitly denies the possibility of arranging either the acids or the bases in any unique order of strength. Soft bases are defined qualitatively as those in which the donor atom is of high polarizability, low electronegativity, easily oxidized, or associated with low-lying vacant orbitals, for example I^-, SCN^-, S^{2-}, $(C_6H_5)_3P$. Hard bases have the opposite properties, and include OH^-, F^-, and many oxy-anions. The distinction between hard and soft acids is along similar lines, though the most useful criterion is their ability to react with the two classes of bases: fortunately it seems possible to arrive at a fairly firm classification on this basis, though there is a borderline category of both acids and bases. Typical hard acids are H^+, cations with a rare gas structure (e.g. Na^+, Ca^{2+}, Al^{3+}) and carbonium ions, while soft acids include Cu^+, Ag^+, Hg^{2+}, and halogen cations. There is no general agreement about the theoretical interpretation of hard-hard and soft-soft interactions, though it has been suggested that they differ in the proportions of ionic and covalent bonding, or in the importance of π-bonding. Nevertheless, a surprising amount of information, concerning both equilibria and rates, can be coordinated on this basis, and further developments will be awaited with interest.

If it is desired to preserve the distinction between proton acids and Lewis acids it seems preferable to confine the term *acid* to the former, and to denote the latter as *electron-acceptors*, which exactly describes their functions. This usage does not automatically imply the qualitative resemblances stressed by G. N. Lewis, but it is quite natural that proton-donors and electron-acceptors should often produce similar effects. The nomenclature of bases offers less difficulty, since a proton-acceptor can always donate an electron-pair to Lewis acids: however, it might be advisable to use the term *base* only in contexts involving the transfer of a proton, since it is only here that the correspondence of acid-base pairs and the quantitative aspects of acid-base strength are applicable. In other contexts the term *electron-donor* is more appropriate.

In conclusion it should be stressed that these questions of nomenclature are concerned only with convenience and consistency, and not with any fundamental differences in interpretation. It is therefore misleading to attach much scientific importance to controversies about acid-base definitions, or to speak of an 'electronic theory' of acids and bases. The chief importance of new definitions lies in their stimulating effect on experimental work. Just as the Brönsted–Lowry definition initiated many investigations of acid-base equilibria and kinetics in different solvents, so the Lewis definition has led to much valuable work on the reactions of acceptor molecules, which will retain its importance even if it is not found convenient to describe them as acids.

GENERAL REFERENCES

G. N. Lewis, 'Acids and Bases', *J. Franklin Inst.*, 1938, **226**, 293.

W. F. Luder, 'The Electronic Theory of Acids and Bases', *Chem. Reviews*, 1940, **27**, 547.

W. F. Luder and S. Zuffanti, *The Electronic Theory of Acids and Bases* (New York, 1946).

R. P. Bell, 'The Use of the Terms Acid and Base', *Chem. Soc. Quarterly Reviews*, 1947, **1**, 113.

'Acids and Bases', collected papers, *J. Chem. Education*, Easton, Pa., 1941.

'More Acids and Bases', *ibid.*, 1944.

Symposium on Hard and Soft Acids and Bases, *Chem. and Eng. News*, **43**, 90, 1965.

R. G. Pearson, *Science*, 1966, 151, 172; *Chemistry in Britain*, 1967, 103.

M. J. Frazer, *New Scientist*, 1967, 662.

acids and bases. The chief importance of acid-base definitions lies in their stimulative effect on experimental work. Just as the Brønsted–Lowry definition stimulated many investigations of acid-base equilibria and kinetics in different solvents, so the Lewis definition has led to much valuable work on the reactions of acceptor molecules, which will retain its importance even if this acid label convention is restricted to Brønsted acids.

GENERAL REFERENCES

G. N. Lewis, *Valence and Structure of Atoms and Molecules*, 1923.

W. F. Luder, *The Electronic Theory of Acids and Bases*, Chem. Reviews, 1940.

W. F. Luder and S. Zuffanti, *The Electronic Theory of Acids and Bases*, Wiley, 1946.

R. P. Bell, *The Use of the Terms Acid and Base*, Quart. Rev. Chem. Soc., 1947, 1, 113.

R. P. Bell, *Acids and Bases*, Methuen, 1952.

A. G. Gaydon, *Dissociation Energies*, Chapman and Hall, 1947.

Symposium on Electrolytes, Discussions Faraday Soc., 1957.

Index